This is Planet Earth

This is Planet Earth

Your ultimate guide to the world we call home

NEW SCIENTIST

First published in Great Britain in 2018 by John Murray Learning
First published in the US in 2018 by Nicholas Brealey Publishing
John Murray Learning and Nicholas Brealey Publishing are companies of Hachette UK

A catalogue record for this book is available from the British Library and the Library of Congress.
UK ISBN 978 1 47 362977 6 / eISBN 978 1 47 362978 3
US ISBN 978 1 47 367038 9 / eISBN 978 1 47 367039 6
1

Typeset by Cenveo® Publisher Services.
Printed and bound in Great Britain by CPI Group (UK) Ltd, Croydon, CR0 4YY.
Hachette UK's policy is to use papers that are natural, renewable and recyclable products and made from wood grown in sustainable forests. The logging and manufacturing processes are expected to conform to the environmental regulations of the country of origin.

John Murray Learning
Carmelite House
50 Victoria Embankment
London, EC4Y 0DZ
www.hodder.co.uk

Nicholas Brealey Publishing
Hachette Book Group
53 State Street
Boston MA 02109
www.nicholasbrealey.com

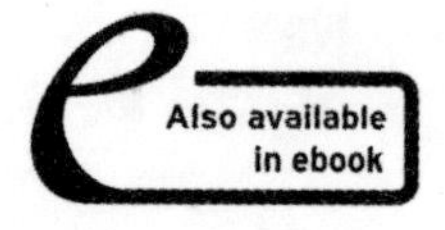

Contents

Series introduction

New Scientist's *Instant Expert* books shine light on the subjects that we all wish we knew more about: topics that challenge, engage enquiring minds and open up a deeper understanding of the world around us. *Instant Expert* books are definitive and accessible entry points for curious readers who want to know how things work and why. Look out for the other titles in the series:

The End of Money
How Your Brain Works
The Quantum World
Where the Universe Came From
How Evolution Explains Everything about Life
Why the Universe Exists
Your Conscious Mind
Machines that Think

Scheduled for publication in 2018:

How Numbers Work
Human Origins
A Journey through the Universe

Contributors

Editor: Jeremy Webb, Editor-at-Large, *New Scientist*

Instant Expert Series Editor: Alison George, Books Editor at *New Scientist*

Guest contributors

David Cromwell writes about ocean circulation in Chapter 7. He is a former researcher at the National Oceanography Centre in Southampton, UK, and is now co-editor of the media analysis website medialens.org.

John Gribbin introduces the structure of the atmosphere in Chapter 6. He is Visiting Fellow in Astronomy at the University of Sussex, UK, and author of numerous books, including *Planet Earth: A Beginner's Guide* (2012).

Susan Hough is a senior seismologist with the Southern California Earthquake Center and a Fellow of the American Geophysical Union. Here, Susan covers what we know about earthquakes and their prediction in Chapter 4.

Jeff Masters focuses on extreme weather in Chapter 6. He is co-founder of the Weather Underground online weather information service, where he is director of meteorology.

David Rimmer is a retired senior lecturer in soil science at Newcastle University, UK. He introduces the section on soils in Chapter 3.

Toby Tyrrell describes the failings of the Gaia hypothesis in Chapter 8. He is a professor of Earth system science at the University of Southampton, UK, and author of *On Gaia: A Critical Investigation of the Relationship between Life and Earth* (2013).

Peter Ward is a professor of biology at the University of Washington in Seattle and author of *The Medea Hypothesis: Is Life on Earth Ultimately Self-Destructive?* (2015). Here, he explores the Gaia hypothesis in Chapter 8.

Thanks also to the following writers:

Anil Ananthaswamy, Colin Barras, Stephen Battersby, Catherine Brahic, Sue Bowler, Stuart Clark, Andy Coghlan, Philip Cohen, Daniel Cossins, Richard Fifield, Linda Geddes, Shannon Hall, Jeff Hecht, Bob Holmes, Joshua Howgego, Ferris Jabr, Victoria Jaggard, Graham Lawton, Michael Le Page, Rick Lovett, Myles McLeod, Michael Marshall, Katia Moscovitch, Rachel Nowak, Sean O'Neill, Stephen Ornes, Jheni Osman, Fred Pearce, Kate Ravilious, Christina Reed, Eugenie Samuel Reich, David Shiga, Colin Stuart, Richard Webb, Sam Wong, Marcus Woo.

Introduction

Earth is an astounding place. You might not think so as you look out of the window and think how ordinary everything seems. But study it closely, as scientists do, and you discover amazing things; things you may find hard to believe. At turns, the planet has been a red-hot blob of molten rock and a giant snowball. The Arctic once basked in tropical temperatures and the entire Mediterranean Sea dried up only to be refilled by the mother of all floods.

Even now we rely on unseen marvels every day. The planet is protected from dangerous ultraviolet rays by one invisible shield, while another keeps it safe from the torrent of energetic particles streaming from the Sun. Our hospitable climate is a wonder in itself, kept in check by the nature of soils, ventilation of the oceans, reflectivity of clouds, the forging and erosion of rock and volcanic eruptions which spout gases that both cool and warm the planet.

And then there's Earth's tour de force: life. You might think that you – as a representative of living things – are unexceptional. But on a cosmic scale that's simply not so. You were born on the only planet in the universe that we know of where life exists – and certainly the only place where organisms read books.

The other thing to bear in mind is that life does not live *on* Earth; it is *part of* Earth. The soil's role in tempering our climate is mediated by microbes. And clouds reflect more sunlight back into space when microbes high in the atmosphere make them whiter. Even some of the rocks of the planet's outer crust are made up of the bodies of long-dead creatures.

Earth is not just a giant boulder flying through space, but a machine where living things interact with geology, water, ice and the atmosphere. Everything is interconnected. *This is Planet Earth* is an introduction to anyone who would like to better understand all these things and how they fit together.

The first two chapters deal with Earth's formation and history, taking us from its first appearance in a swirling cloud of gas around the young Sun to the tsunami that cut off Britain from the rest of Europe. Chapters 3 looks at Earth's structure. We venture from the surface, with its life-giving covering of soil, right down to its hard iron heart. Chapters 4 and 5 develop the idea of plate tectonics, which has enabled us to understand more about such things as earthquakes, the planet's thermostat and the future wanderings of our continents.

Chapters 6, 7 and 8 explore different 'spheres' of the planet, starting with the atmosphere. We spend most time in the lowest breathable layer where life exists and weather happens, but also venture all the way up to the edge of space. For the hydrosphere we dive into the oceans to examine the gigantic pumps that drive global currents. Finally, in the biosphere, we search for clues to how life got started and how it has been influencing the planet ever since.

To close, we focus on humanity's influence on Earth's systems. Chapter 9 introduces the Anthropocene, a proposed geological epoch designed to recognize our profound impact on the globe. Chapter 10 deals with the biggest threat we know of today to the existence of life – climate change. We check what we know, where the gaps are in our knowledge, and ask whether we can fix things.

Hopefully, this book will change your perspective on Earth so you will never see it as ordinary again. Who knows, you may even be astounded.

Jeremy Webb, Editor

I
Formative years

The Earth and Moon were born from chaos. The heat and violence of the early solar system have conspired to shroud much of Earth's early years in mystery. What, then, do know and not know about how it developed into the dynamic planet we know today, a place fit for life to evolve?

A home unique in the universe

Very occasionally, routine events produce exceptional results. That's what happened 4.6 billion years ago, on a minor arm of an unremarkable spiral galaxy.

A vast cloud of gas and dust began to collapse into a dense ball of matter. As gravity pulled more and more material towards it, the temperature and pressure at its core increased to the point where nuclear fusion kicked in. That released vast quantities of energy and marked the creation of a star.

What started the process off, we don't know but it had happened countless times before and the star itself was certainly nothing special.

As the newborn star began to spin, smaller bodies started to coalesce in orbit around it. Gas molecules and dust particles fused to form objects the size of rocks; which collided to create boulders, then 'planetesimals'. Their increasing gravity pulled in still more matter to create hot, molten versions of the planets we know today.

Eight planets formed and on the third one from the star something truly remarkable happened. The right conditions enabled life to emerge and flourish. Eventually, intelligent life evolved in the form of beings capable of asking how their planet had formed and how it came to nurture life. They called their celestial neighbourhood the solar system, named the star 'Sun' and their planet 'Earth'.

Mysterious beginnings

That's the big picture, at least. 'Time zero' for the solar system is generally agreed to be 4.567 billion years ago, and by 4.55 billion years ago, about 65 per cent of Earth had assembled.

The early solar system was an energetic, dynamic place. For its first few hundred million years, collisions were common and Earth was subjected to some pretty rough treatment. About 4.53 billion years ago, just as the paint was drying on the infant Earth, disaster struck. It was dealt a glancing blow by an object the size of Mars. The impact threw debris into orbit to form the Moon, and the energy of the collision melted Earth's upper layers, erasing any previous geological record. Vaporized silicon that didn't make it to the Moon condensed and fell back as lava rain, depositing a sea of molten rock. Earth eventually melted to its core, and the process of forming a solid surface began all over again.

This version of the Moon's creation is not the only one, as you will see later. Yet it seems certain that the violence carried on, ending only with a sustained pummelling between 4.1 and 3.8 billion years ago in what is known today as the late heavy bombardment. Once again, the ferocity and length of this episode are still being debated.

The sheer violence of these events is one reason why there is a yawning chasm in our knowledge of Earth's first 500 million years, an aeon known as the Hadean, after Hades, the ancient Greek god of the underworld. With little to go on scientists make up stories that best fit the evidence they have; evidence that comes from our knowledge of physics and chemistry, the results of hands-on experiments, observations of other astronomical objects and computer simulations.

Research to answer many of our questions is under way right now and new findings, observations and models are being made all the time. What we think we know is constantly being challenged by that new evidence. And so scientists' stories change.

Among our unanswered questions is how Earth gained so much water. Being close to the Sun, it was probably too hot for water to simply condense out of the gas cloud as the planet formed. In any case, water that had collected would probably have evaporated during the titanic collision that formed the moon. One possible explanation is that the water arrived later, delivered by icy comets and asteroids from the outer solar system during the late heavy bombardment.

Then there's the question of when Earth gained its crust. Today that crust is composed almost exclusively of rocks no older than 3.8 billion years, so traces of the hellish Hadean are thin on the ground. Of the ancient rocks that remain, most have been modified by heat and pressure. The good news is that tiny resilient crystals, called zircons, may be seriously old and are yielding important information. Combined with ever-improving methods of microanalysis, these may yet rewrite the story of the early Earth.

There is one other way we can learn more about the Hadean. Mineral prospecting on the Moon and Mars could also reveal what Earth was like before the great impact. Unlike Earth, neither of those worlds has re-melted, so there is a much greater chance of finding truly ancient rocks on their surfaces. We may even hit the geological jackpot and find a piece of the Hadean Earth that was blasted into space by an asteroid impact, and which subsequently landed on the Moon or Mars.

With this overview of Earth's earliest years in mind, let's drill deeper into the issues that are keeping earth scientists, astrophysicists and palaeobiologists up at night.

Our enigmatic moon

Accounting for the Moon's origin has always been a problem. It is just too big. No other planet in our solar system has a satellite that is proportionally so large: it is over one-quarter of

Earth's diameter. Such a body could not have been captured in passing, as other planets are thought to have snared their smaller satellites. In 1879 George Darwin, the astronomer son of Charles, proposed a different idea. He suggested that the early Earth spun so quickly it fell apart, spitting a bit of itself into space.

That idea was popular for a time, but fell foul of planetary dynamicists in the early twentieth century, who found that the numbers just did not add up. For Earth's outwardly directed centrifugal force to overwhelm the inwardly acting gravitational force and break the planet apart, it would have had to be rotating ridiculously fast, at about once every two hours.

Darwin's idea has been replaced by the giant impact hypothesis or 'big splat' – that glancing blow from a Mars-sized object (see Figure 1.1). In the maelstrom of colliding objects in the early solar system it is perfectly reasonable to expect huge impacts in its latter stages.

However, the big splat itself could be quashed by an analysis of Moon rocks brought back by the Apollo astronauts. According to the giant impact hypothesis, some of the Apollo rocks should have come from the object that crashed into our planet, yet analysis shows that their oxygen, chromium, potassium and silicon isotopes are indistinguishable from Earth's. Added to this, several samples thought to come from the Moon's crust contained water. In the aftermath of the big splat, the heat generated should have melted the rocks and driven off the water.

Still, all is not lost. Ignoring the highly speculative idea that a natural nuclear reactor within Earth exploded and blew a portion of the planet into space, further modelling shows that there are types of collision that overcome these criticisms.

Matija Cuk of the SETI Institute in Mountainview, California and Sarah Stewart at the University of California, Davis

FIGURE 1.1 Did a 'big splat' create our moon?

found that, if Earth spun faster than assumed in the past, it would have needed less of a smack to spit out the Moon. Instead of a Mars-sized impactor, one with just half the mass could have hit Earth at a steeper angle, burying itself deep inside our world. Cuk and Stewart's computer simulations show that such an event would provide just enough energy to explode a plume of solely Earth rocks into orbit – providing a moon isotopically indistinguishable from Earth.

A different sort of 'giant impact lite' has been proposed by planetary scientist Robin Canup of the Southwest Research Institute in Boulder, Colorado. She envisages two planets, each about half the size of Earth, colliding slowly. In the ensuing coalescence that gave birth to our planet, the Moon was formed from the leftovers, ensuring that both bodies were made from the same ingredients.

Was it really hell on Earth?

The chaos of the early solar system certainly sent objects hurtling around. Some came our way – one look at craters on the Moon gives some idea of the scale of the battering. We don't see this damage on Earth because it has been eroded by wind, rain and plant life.

Those rocks brought back to Earth by the Apollo astronauts suggest that the most intense battering took place during the late heavy bombardment. Before and during this battering, it's widely thought that our planet was a molten hellhole, far too hot, dry and hostile for life to get going. Not until the bombardment ended during the Archaean aeon (see Chapter 2) did conditions improve enough for life to gain a foothold.

In recent years, however, the late heavy bombardment has started to be questioned on several fronts. Most radically, some scientists argue that the spike in lunar bombardment may be an artefact of how the Apollo evidence was gathered.

The Apollo samples came from different sites on the Moon. Even so, researchers have pointed out that they may all have come from a single event – the impact or impacts that formed the Imbrium Basin, one of the large, dark patches that make up the 'Man in the Moon'. Rocky fragments from this event could have contaminated disparate parts of the lunar surface, meaning that what at first looked like a host of simultaneous impacts might have been only a handful. If the impacts were more smeared out, early Earth would not have been so hellish as first assumed.

Evidence of a milder Hadean is also coming from other sources, including the tiniest of witnesses. Zircons are tough crystals of zirconium silicate, often a millimetre or less in length and among the oldest objects on Earth. They can survive being baked at 1600 °C and washed down the course of an entire river without chipping. Most importantly for geologists, they can survive under tonnes of sediment without undergoing metamorphosis or melting, as other materials do.

Zircons are ubiquitous on Earth. They are found in almost all granites, which form when rock is re-melted within the Earth before rising and cooling. As granite solidifies, any zirconium in the melt snatches up silicate and crystallizes out as zircons. They are also common in sedimentary rocks once erosion has freed them from their original granites.

The zircons raising doubt about the Hadean come from rocks in the Jack Hills of Western Australia, which themselves date back some 3.7 billion years. When zircons from here were dated – by comparing the proportions of uranium with its radioactive decay products (see Chapter 2) – they turned out to be even older. One dated from 4.4 billion years ago, which suggests that solid material existed on the surface of Earth a mere 200 million years after its birth.

What's more, researchers found 'inclusions' – traces of quartz, mica and feldspar – trapped inside the zircons. These suggest that the zircons had formed from melted, metamorphosed sediments that might initially have been similar to wet mud or clay. Analyses also found the ancient zircons to have high concentrations of the isotope oxygen-18. Rocks that form at a low temperature in wet conditions tend to absorb more oxygen-18 than other rocks.

Measurements of other elements in the zircons support this finding. It now seems that, far from being a 'magma ocean' with no atmosphere, Earth 4.4 billion years ago was solid, coolish and wet. Ponds, pools and oceans need a solid surface to sit on, suggesting that a crust was in place early on, and if there was liquid water then there had to be a thick atmosphere: otherwise the water would have boiled off. Hell on Earth suddenly looks rather balmy.

When did life get started?

The idea that Earth in the Hadean was cooler and wetter than previously thought is reflected in recent, though controversial, new evidence for the earliest life.

The earliest reliable evidence of life comes from a fossilized beach in the Pilbara region of Western Australia, north of the Jack Hills. That was dated to 3.43 billion years ago. Chemical signatures in even older rocks in south-western Greenland suggest life may have been present 3.8 billion years ago, though this evidence is contentious.

In 2015 Elizabeth Bell and Mark Harrison from University of California, Los Angeles and colleagues announced that they had found carbon with an organic-like signature sealed within a zircon crystal. The team had analysed more than 10,000 zircons from the Hadean and Archaean aeons. In one Hadean crystal from the Jack Hills (see Figure 1.2) they found tiny flecks, or inclusions, of graphite, which must have been incorporated into the zircon when it formed some 4.1 billion years ago.

The researchers analysed carbon isotopes in two of the graphite flecks, and found them to have a high carbon-12 to carbon-13 isotope ratio. This is a characteristic of biological origin because living organisms preferentially absorb carbon-12.

FIGURE 1.2 Australia's Jack Hills hark back to Earth's violent youth, creating a window on the Hadean aeon.

The general chemical make-up of the zircon crystals suggests that the magma they cooled from was generated by the melting of a mud-rich sediment, which is the sort of environment in which organic remains might accumulate.

Then, in 2017, Matthew Dodd at University College London and co-workers reported evidence of even earlier life in rocks dating back to the Hadean, or close to it. These rocks, collected from the Nuvvuagittuq Greenstone Belt on the eastern shore of Hudson Bay in northern Quebec, Canada, are at least 3.75 billion years old, and some geologists argue they are 4.28 billion years old, making them only slightly younger than the planet itself.

Like all such rocks, they have been heavily altered. At some point, they spent time deep inside Earth, where temperatures above 50 °C and extreme pressures baked and deformed them.

But geologists can still read clues that suggest they formed at the bottom of Earth's early oceans: they seem to preserve evidence of ancient deep-sea hydrothermal vents.

Dodd's evidence stems from iron-rich rocks originally formed around relatively cool (less than 160 °C) undersea vents. These rocks contain microscopic tubes and filaments made of iron oxide that are very similar to structures formed by bacteria living today in mat-like colonies around deep-sea hydrothermal vents.

What's more, material close to the filaments contains a high carbon-12 to carbon-13 ratio, which implies organic origins. Some of that carbon is inside crystals of phosphorus-rich minerals, which also hints at early biology because phosphorus released as organisms decay can be incorporated into minerals.

If confirmed, the findings would potentially push the origin of life back to 4.29 billion years ago, suggesting that Earth was inhabited astonishingly early, even before the late heavy bombardment. More than that, it would show that life got going around deep-sea vents where there is little or no sunlight, so organisms would have had to derive their energy from geothermal processes. This would help to bring the geological evidence in line with findings from genetic and biochemical studies hinting that life emerged in deep hydrothermal areas – and not in shallow, sun-drenched environments where most early fossils have been found (see Chapter 8).

The case for when life emerged is not closed yet, however. Neither study is conclusive. Harrison's team acknowledges that there are also inorganic ways that isotopically light carbon could have accumulated in Hadean environments. And it would be extraordinary if the fragile, microscopic structures found by Dodd survived in rocks that have been subjected to high temperatures and pressures deep underground.

A spanner in the works

Just when you thought the story of early Earth was emerging from the fog, along come results that force you to think again.

Donald Lowe of Stanford University in California and his colleagues have spent 40 years studying a patch of ancient rocks in eastern South Africa called the Barberton Belt. More than 25 years ago they found four layers of spherical particles, which seemed to have condensed from clouds of vaporized rock. Lowe says they are the traces of four major meteorite impacts, dating from between 3.5 and 3.2 billion years ago. In 2014 his team described another four layers from the same period. He argues that eight major impacts within 250 million years bolsters the case that the late heavy bombardment was longer than most researchers think, and tapered off only about 3 billion years ago.

The impacts were on a scale beyond anything Earth has experienced since the dawn of complex animals. The asteroid believed to have finished off the dinosaurs left a layer of spherules a few millimetres thick. Lowe's layers are 30 to 40 centimetres thick, suggesting that the asteroids were at least 20 kilometres across and possibly more than 70. Impacts from such objects would wipe out most animals and plants if they happened today, but back then all life was single-celled and aquatic.

Lowe reckons a very large impact could have heated the atmosphere to hundreds of degrees Celsius and evaporated the top 100 metres of the ocean. Microbes on the opposite side of the planet might have been able to ride out large waves and the rain of hot rocks. But one group likely to have fared especially badly was photosynthetic bacteria because they would have had to live near the ocean surface where there was plenty of light.

Water, water everywhere

If water existed on Earth during the Hadean, it raises the question of where it came from. It's true that the early Sun would have been weaker than it is today, but even so it would surely have vaporized any ice in the primordial cloud that gave birth to Earth, which is why planetary scientists assumed that water was delivered later by celestial messengers – comets or meteorites. Recent evidence, however, suggests this is not the case.

The key to figuring out where Earth's water came from lies in the ratio of concentrations of two hydrogen isotopes: deuterium, also called heavy hydrogen, and normal hydrogen. This ratio differs depending on the source of the water. So comparing the ratio in, say, meteorites with that in Earth's oldest water should reveal if their H_2O had a common origin.

Over deep time, the oceans have almost certainly lost some of the lighter hydrogen isotope, so instead researchers turned to water in ancient volcanic basalt rocks from Baffin Island in the Canadian Arctic. These rocks contain tiny glassy inclusions that appear to have formed in Earth's mantle, the layer below the crust, 4.5 billion years ago. This would make them almost as old as the planet itself, and locked inside them are hydrogen molecules from water of the same age.

These inclusions contain surprisingly little deuterium: their ratio is nearly 22 per cent less than in seawater today. The result points to a source that's very deuterium-poor, which probably rules out meteorites because their isotope ratio is usually higher. Instead, the ratio suggests that the water must have originated in the cloud from which the Sun and planets condensed.

This conclusion supports theoretical studies which show that water molecules would have clung tightly to the coalescing dust particles even in the hot conditions of Earth's formation.

In recent years scientists have also found that Earth has far more water than expected deep beneath its surface. Some of this may have migrated upwards to the surface. The internal reservoir is estimated to contain up to three times the volume of all the oceans. It resides in a blue rock called ringwoodite, a magnesium silicate mineral that forms at temperatures and pressures found 600 kilometres down in the mantle.

Other studies suggest that this reservoir may not even be the deepest. Hydroxyl ions, usually a sure sign of water, have been found in inclusions within diamonds coughed up in lava. These form at depths of around 1000 kilometres, suggesting that water may be cycling down into the deep mantle.

Restless crust

We now know that Earth is covered by several large, rigid plates that are constantly moving and rubbing against each other – a process known as plate tectonics (see Chapter 4). This process constantly recycles Earth's rocks, and without it the planet would not have a stable climate or the oil and mineral deposits we depend upon.

Earth is the only planet we know that has plate tectonics. Why should this be? Models show that for tectonics to get going a planet has to be just the right size: too small and its lithosphere – the solid part of the crust and upper mantle – is too thick. Too big and its gravitational field squeezes any plates together, holding them tightly in place. Other conditions have to be right too: the rocks making up the planet should be not too hot, not too cold, not too wet and not too dry.

Our thinking about Earth's first crust is based upon the processes we see today. In the here and now, oceanic crust forms at mid-ocean ridges, where molten rock from the mantle flows up and spreads out. The resulting rock is rich in hard, black basalt – the rock that makes the volcanic islands of Hawaii.

Continental crust is different. It tends to be made of rocks such as granite, which form when basalt sinks, melts and reforms. In the process it gets enriched with silica, aluminium and lighter metals. Granite is less dense than basalt, so it floats higher on the mantle than oceanic crust.

Where these two types of crust meet, for example at the rims of many ocean basins, the cold, dense oceanic floor, together with plenty of water, mud and ooze, dives under the lighter continental crust and into the mantle – a process called subduction.

So much for the present. The big question about the past is when and how did the lithosphere become cracked in such a way that the first piece of crust dived down beneath another?

Because of the turmoil of the Hadean, evidence is thin on the ground. But geochemical research published in 2016, on zircons and rocks from north-western Canada dated to 4 billion years ago, suggests that Earth's skin at the time contained no continental crust, but was more like basaltic oceanic crust. If this is confirmed, then plate tectonics as we know it did not take place during the Hadean.

One oft-quoted description sees Earth, for perhaps its first 2 billion years, as having a thin skin of basaltic crust covered with water and punctuated with chains of volcanoes as just about the only land. These would have belched out water vapour and gases such as carbon dioxide, sulfur dioxide and hydrogen chloride to create a thick atmosphere.

How the first cracks appeared in this crust is still moot. A hot plume of material in the mantle could have punctured the crust to make the first hole, or perhaps an asteroid or comet impact was the trigger, piercing the surface layer and setting up a chain of events that created the first moving tectonic plates.

When did this happen? One estimate stems from research into ophiolites, rare slivers of ancient oceanic crust which, instead of being pushed down into the mantle, are raised up on top of continental crust at a subduction zone. A 2007 study dated a sample of what is thought to be an ophiolite in Greenland to 3.8 billion years ago – the oldest suggestion of plate tectonics yet.

Watching other worlds

New insight into how Earth got its crust comes from observing Jupiter's tiny moon, Io. Some scientists think that the very early Earth was some sort of magma ocean – all the planets seem to have passed through this sort of phase. But how did Earth move on to plate tectonics?

Io is covered in active volcanoes that transport heat from its interior to its surface without plate tectonics. It loses heat through heat pipes, a system of volcanic plumbing that shuttles hot molten rock, or magma, to the surface through relatively narrow channels. The lava cools as it spreads, forming a new layer of crust that is later covered by fresh eruptions. Over time, the heat pipes create a thick upper layer of crust, which on Io is strong enough to support mountains more than 20 kilometres high.

William Moore of Hampton University in Virginia and Alexander Webb of Louisiana State University in Baton Rouge ran simulations of Io-like heat pipes on early Earth to see what kinds of rocks were created and how the crust would have behaved. They compared the results to the most ancient known rocks on Earth, including 3-billion-year-old diamonds and 4.3-billion-year-old zircons.

The pair found enough correlations to suggest that, until about 3.2 billion years ago, Earth released its excess heat through Io-like heat pipes scattered over a few regions of an otherwise barren surface. Eventually, they propose, Earth cooled so much that the heat pipes shut down, which allowed stress to build in the hot mantle trapped beneath the lid of crust. According to this version of events, increasing stresses in the convecting mantle broke the outer skin and plate tectonics began.

Our changing planet

In the past 4.5 billion years, everything about our planet – from its atmosphere to its core – has undergone extraordinary transformations. The dates here are fluid because research is continually uncovering new evidence.

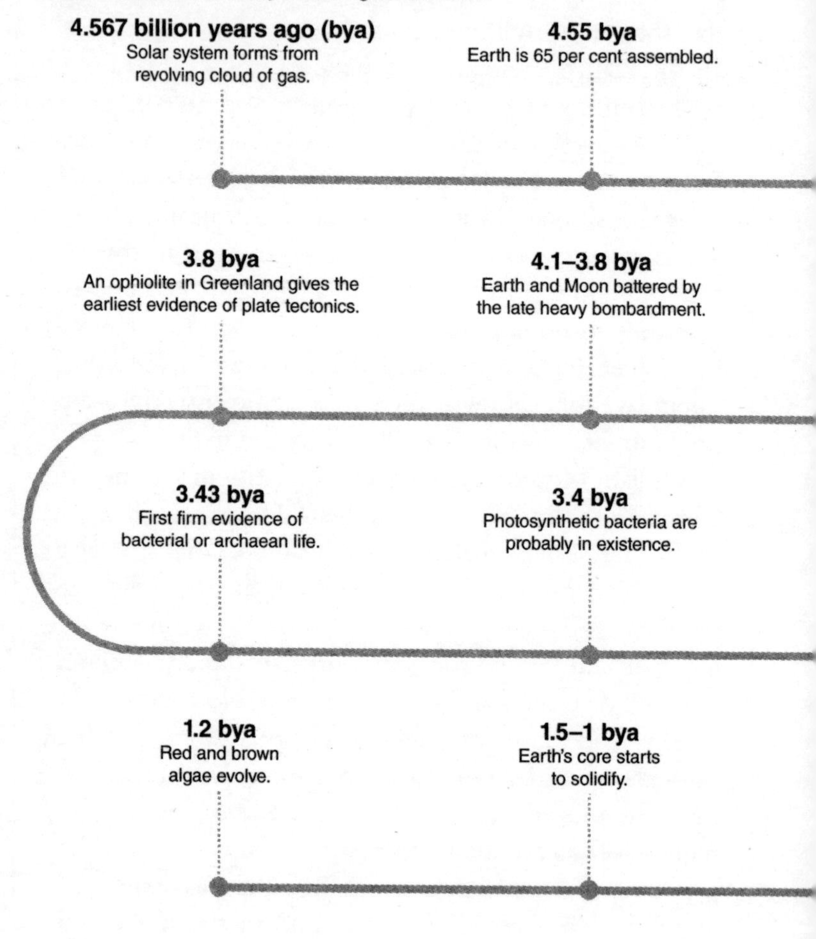

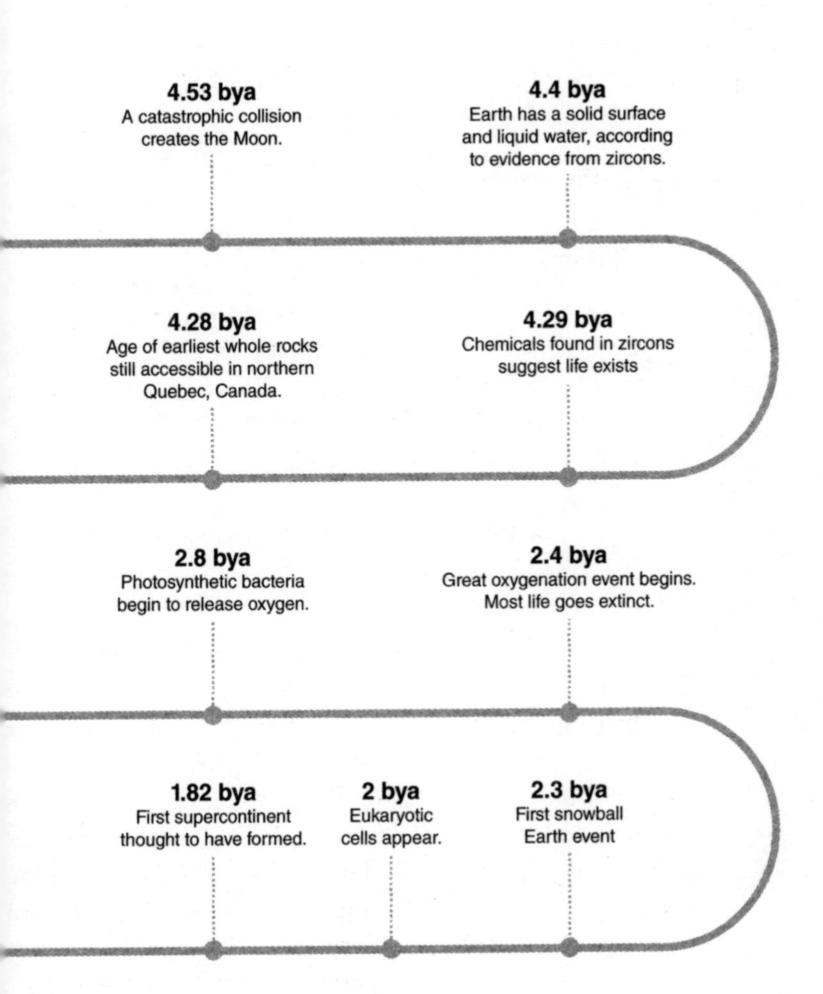
4.53 bya
A catastrophic collision creates the Moon.
4.4 bya
Earth has a solid surface and liquid water, according to evidence from zircons.
4.28 bya
Age of earliest whole rocks still accessible in northern Quebec, Canada.
4.29 bya
Chemicals found in zircons suggest life exists
2.8 bya
Photosynthetic bacteria begin to release oxygen.
2.4 bya
Great oxygenation event begins. Most life goes extinct.
1.82 bya
First supercontinent thought to have formed.
2 bya
Eukaryotic cells appear.
2.3 bya
First snowball Earth event

Our changing planet

(continued)

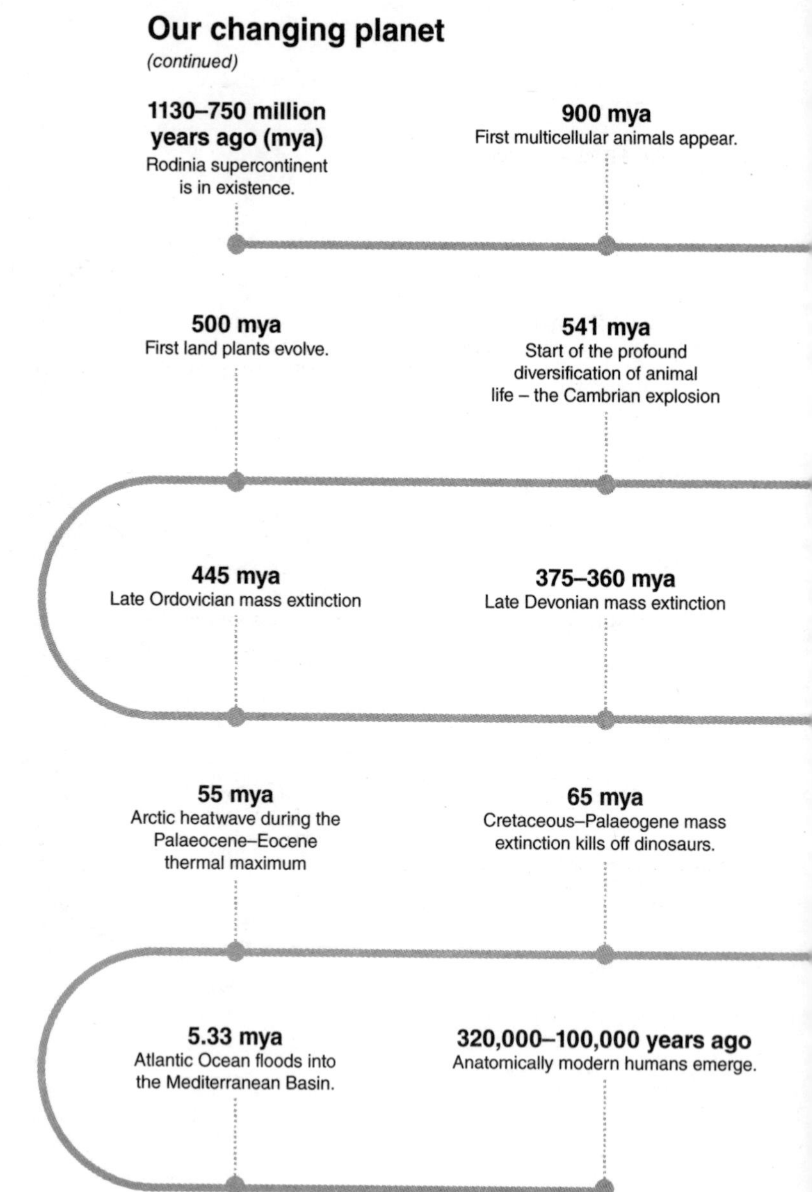

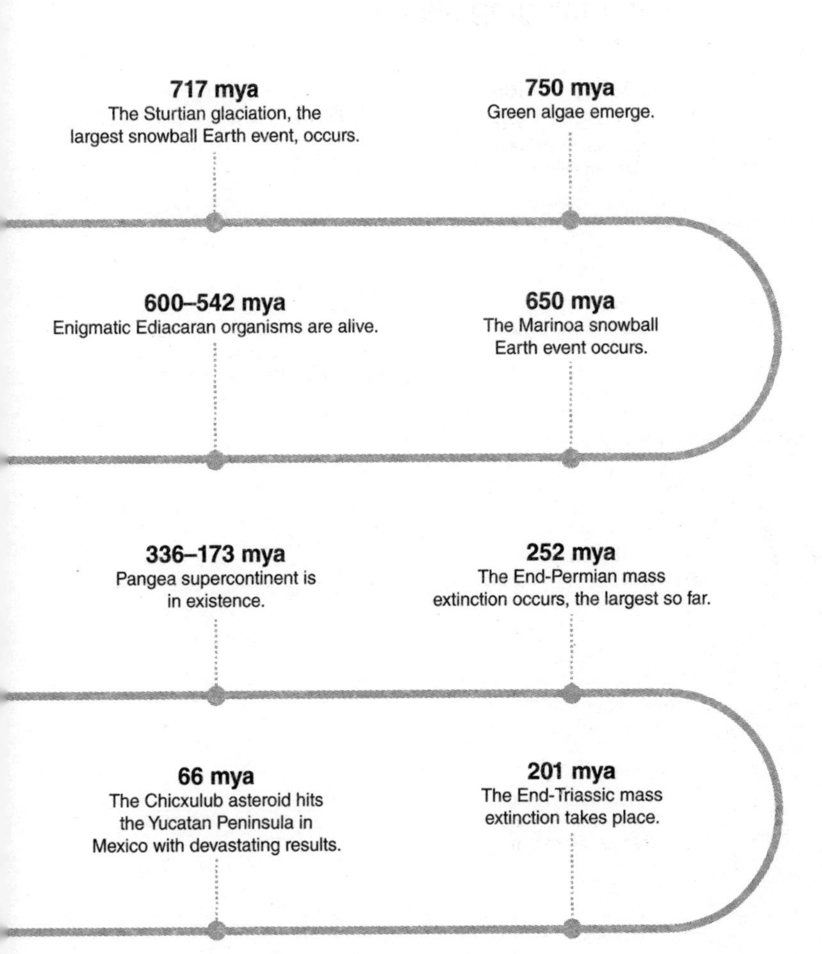
750 mya
Green algae emerge.
717 mya
The Sturtian glaciation, the largest snowball Earth event, occurs.
650 mya
The Marinoa snowball Earth event occurs.
600–542 mya
Enigmatic Ediacaran organisms are alive.
336–173 mya
Pangea supercontinent is in existence.
252 mya
The End-Permian mass extinction occurs, the largest so far.
201 mya
The End-Triassic mass extinction takes place.
66 mya
The Chicxulub asteroid hits the Yucatan Peninsula in Mexico with devastating results.

2
The long sweep of time

The early years of Earth were punctuated by astonishing events, though it's easy to forget that they unfolded over hundreds of millions of years. Indeed, the slow pace at which most geological processes unfold on Earth can be difficult to appreciate. Here we take a look at deep time, how we measure it and check out a few of the landmark episodes that have punctuated Earth's long journey through time.

Time goes deep

In June 1788 Scottish geologist James Hutton took his colleagues John Playfair and James Hall to Siccar Point on the Berwickshire coast. To unenlightened eyes, the rocky promontory would have appeared eternal and unchanging. But Hutton knew better. He had concluded that layers visible in the rocks had formed beneath the sea from the remnants of older rocks and had been raised back up.

Hutton realized the sequence of events he could read in the rocks spoke of incredibly slow changes occurring over mind-expanding stretches of time. The 'angular unconformity' of rock layers of varying types and orientations could only have formed over tens of millions of years (see Figure 2.1). These observations were crucial to his theory of Earth's gradual evolution and the revolutionary concept of deep time.

That time might be counted in millions of years was revolutionary because little more than a century earlier the Primate of All Ireland, Archbishop James Ussher, had used the Bible and other sources to pinpoint the date of creation to Sunday, 23 October 4004 BCE. Isaac Newton disagreed: he thought the year was 3988 BCE.

Then, as now, deep time went against the grain of common sense. After all, says John McNeill, an environmental historian at Georgetown University in Washington, DC, measuring things against a human lifespan is a normal and natural way to think. Through the heroic efforts of Hutton and many after him, we now know that Earth is around 4.56 billion years old, an almost inconceivable age.

Deep time has been central to the development of the historical sciences – geology, evolutionary biology and cosmology – and remains so. Without it, we can't appreciate that some processes,

FIGURE 2.1 Rock strata tell tales about their age, how they were forged and what's happened to them since.

whether the weathering of rocks, the building of mountains or the evolution of species, generally take place on timescales so slow as to be inappreciable within a human lifespan.

The trip to Siccar Point prompted Playfair to write: 'The mind seemed to grow giddy by looking so far into the abyss of time.' If odd glimpses of deep time are hard to deal with – marine fossils high above the water line, for example – they can also be liberating. As McNeill puts it: 'What it means to me is that we are all part of an unimaginably long chain of being, both human and non-human, and our own travails don't amount to a hill of beans.'

Deciphering the rocky record

Geologists can trace the history of Earth back nearly all of its 4.56 billion years. They divide this vast span into intervals that form the basic yardsticks of geological time. Early geologists named these intervals on the basis of the rocks formed within them but without knowing when they started and ended. Adding dates to create a timescale fell to later generations.

On Earth, wind and water deposit bits of older rocks and other detritus to be compressed into sedimentary rocks – part of the planet's great rock recycling effort. Early geologists found that these sedimentary rocks formed in layers, where new material builds up on top of old. William Smith, an English surveyor of coal deposits and canals around the start of the nineteenth century, recognized that these layers – or strata – formed regular patterns in rocks and that the deeper the rock lay underground, the older it was.

It's not quite as easy as that, of course. Sequences of sedimentary rocks can include breaks, or discontinuities, where rocks did not form or were eroded away. And not all types of rocks generate neat layers. Igneous rocks form when molten rock – magma or lava – solidifies, and if they do create layers they are often irregular. Metamorphic rocks form when heat and pressure within Earth modify existing rocks so much that new minerals form. Movements in Earth's crust can even fold old sedimentary layers so they appear on top of younger ones.

Geologists, alert to such quirks, build the stratigraphic scale by studying and correlating strata over wide areas. To establish worldwide correlations, they use events that leave a recognizable mark at exactly the same time over large areas. One example is the iridium-rich dust from the asteroid impact implicated in killing off the dinosaurs at the end of the Cretaceous (see Figure 2.2). Large volcanic eruptions scattering ash over large areas are more frequent, and the ash often bears a distinctive chemical signature.

Fossils are also useful markers. The ideal index fossil for dating is one that is common, and evolved rapidly. For example, trilobites are widely used for dating the Cambrian period. There were many species of trilobites, some evolving before others, some dying out earlier. Any rock that contains a particular combination of these fossils must have formed in the particular interval of time when those creatures lived. The disappearance of many types of fossil after mass extinctions provides other definitive signals.

Another modern marker is the weak magnetic field held by some rocks. The remanent magnetism of an igneous rock shows the direction of Earth's magnetic field at the time it solidified. Fortunately for geologists, the planet's north and south poles have swapped position many times in its history. The periods of normal and reversed magnetism vary in length, so they form a distinctive pattern that can be used to correlate sequences of rocks from around the world.

Using such markers, geologists can put sequences of rocks in the right place in the stratigraphic scale. They also compare sequences of rocks of the same age around the world

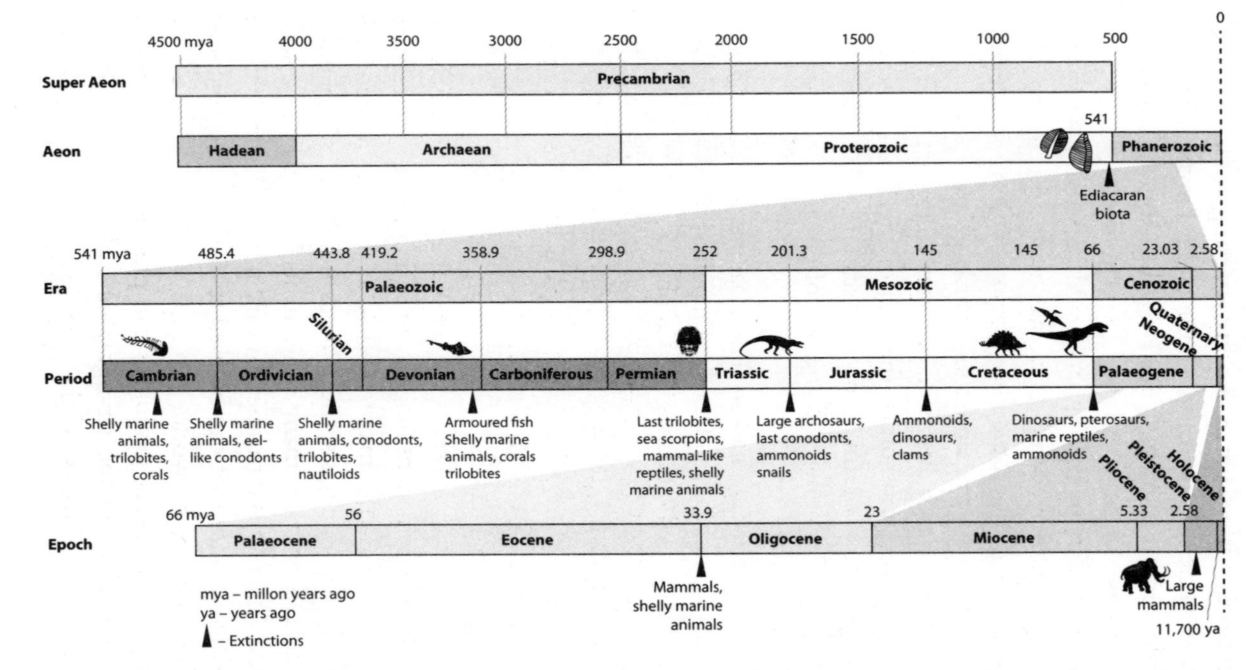

FIGURE 2.2 The geological timescale is divided into a hierarchy of different phases. Boundaries are defined by a variety of markers, including the extinctions of species.

and select the one that is the most complete, with clearly defined layering and the most useful fossils. For any boundary between geological periods, this becomes the Global Boundary Stratotype Section and Point or 'golden spike' for that boundary (see Chapter 9).

Radiometric dating

Finding a place in the stratigraphic scale for a sequence of rocks is one thing. Giving it a chronological date is another. The best tool for finding the age of rocks is radiometric dating, based on the decay of radioactive isotopes.

The natural clock geologists use is the 'half-life' of these isotopes, the time it takes for half the nuclei in a sample to decay. If you start with a million atoms of an isotope with a half-life of 100 years, only half will remain after 100 years, and half of that half – one-quarter – after another hundred years. Thus, as time passes, a rock will contain less of the original radioactive isotope and more of the isotopes it decays into.

Geological dating needs radioactive isotopes that are reasonably common and have very long half-lives, comparable to the ages of the rocks themselves. One choice is the family of uranium and thorium isotopes that eventually decay to lead. The most common is uranium-238, with a 4.5-billion-year half-life, which decays to lead-206. Good radiometric dating gives ages to within 0.25 per cent, a margin of 250,000 years for a date 100 million years ago. These dates have given geologists a solid chronological calibration for points on the geological timescale.

Powerful though it is, radiometric dating has important limitations, notably that it cannot give a direct date for the deposition of sedimentary rock. But all is not lost because a lava flow will be younger than the rocks it flows over and older than the sedimentary rocks that settle above it. Lava is an igneous rock, suitable for radiometric dating, so it can produce a chronological age.

The dating hierarchy

Early geologists chose the first fossil evidence of life to mark the start of the Cambrian. They called the entirety of earlier time 'Precambrian'. Later, geologists found earlier signs of life but left the Precambrian-Cambrian boundary in the same place on the timescale. It now marks the appearance of hard-shelled animals.

The Precambrian is classed as a superaeon and contains three aeons, the longest common divisions in the timescale. These are the Hadean, the Archean (meaning 'origin') and the Proterozoic ('early life'). The most recent – the present – eon started 541 million years ago and is known as the Phanerozoic ('visible life').

The Phanerozoic is subdivided, because geologists know much more about it than earlier aeons, into three 'eras': the Palaeozoic ('early life'), Mesozoic ('middle life'), and Cenozoic ('recent life'). Early geologists made those divisions based on changes they saw in the fossil record. We now know that several distinct dividing lines mark global extinction events in which large numbers of species vanished in a geological instant.

By far the worst of these came at the end of the Palaeozoic era – the End-Permian extinction. It killed off an astonishing 80 to 90 per cent of land and marine species. But it made way for the dinosaurs that evolved to dominate the planet in the Mesozoic era. They gave up the ghost – with the exception of birds – at the end of the Mesozoic era 65 million years ago, after an asteroid smashed into the planet. This in turn allowed the first mammals, including primates, to flourish.

Eras are further broken down into periods, which are mostly named after characteristic rocks or the areas in which they appear. British geologists were central to this exercise: the Cambrian period was named for Cambria, the Roman name for Wales; the Ordovician and Silurian were named for Welsh tribes. Rocks characteristic of the Devonian period were found, you guessed it, in Devon.

Periods last for tens of millions of years, so geologists split them into still smaller divisions called epochs. Most of these subdivisions are important only to specialists, except in the Cenozoic era, where epoch names are widely used because of the better resolution of dates in younger rocks. The epochs become shorter towards the present, reflecting our greater knowledge of recent events. The Holocene spans only the past 11,000 years, essentially the time since the last great ice sheets melted in Europe and North America.

Earth through the aeons

Earth has a long past, during which it has gone through some astonishing transformations. And if you thought Earth today has spectacular natural features, they are nothing compared to those of the past. Here are seven wonders of our forgotten world.

Rivers deep, mountains high 1 billion years ago

A billion years ago, when Earth's landmasses were fused in a supercontinent called Rodinia (see Chapter 5), the world was totally unfamiliar. All life then was unicellular and entirely aquatic, so Rodinia's vast expanse was completely barren. But what it lacked in biological richness it made up for in vast river systems and mountain ranges (see Figure 2.3).

As the continental building blocks that made Rodinia crashed together about 1.2 billion years ago, large portions of crust were lifted up, much as the ongoing collision between the Indian and Eurasian plates is generating the Himalayas now. They were probably no higher than mountains today. According to geologist David Rowley of the University of Chicago, weathering erodes away mountains as they form, and gravity dictates how much load Earth's crust can bear without buckling. Account for that, and Mount Everest is about as high as a mountain can be.

But, by another measure, Rodinia's mountains are extraordinary. Imagine taking the Andes, Rockies, Himalayas, Alps, Atlas and Urals and stringing them together end to end, and you're getting close to the length of Rodinia's principal mountain chain. The range stretched across the entire

FIGURE 2.3 Braided river systems like those found in the Arctic once covered the whole of Rodinia.

supercontinent, maybe 15 to 20,000 kilometres, says Robert Rainbird of the Geological Survey of Canada in Ottawa. Its eroded remains can still be found across North America and Europe, including parts of the Appalachians and the Highlands of Scotland.

And just as mountain ranges like the Andes and the Himalayas give rise to great rivers today, so, too, did those of Rodinia – but with a big difference. There was no vegetation to constrain the rivers so they would have just flowed unconstrained across the barren landscape, Rainbird says. Similar river systems, characteristically braided into many smaller channels, exist in the vegetation-free high Arctic today, but on Rodinia they were vastly bigger. They would have been far longer and wider than the Amazon, and fed inland seas far bigger than anything we have on Earth now, says Rainbird. He and his colleagues have found sediments from the Rodinian mountains 3000 kilometres away on the other side of North America, as well as in India, Antarctica, Scandinavia and Siberia.

Rodinia began to break apart about 750 million years ago, splitting its vast mountain range into pieces. By the time the landmasses reassembled into the next supercontinent, Pangaea, around 300 million years ago, the land was covered in vegetation. So, while Pangaea might also have been home to long mountain ranges, the great rivers of Rodinia are possibly unique in Earth's history.

Snowball Earth 700 million years ago

These days, if you want to see a glacier near the equator, you must scale the rarefied heights of Mount Kenya or the Ecuadorian Andes. Around 700 million years ago it was less of an effort. In fact, you'd struggle to find somewhere that wasn't frozen over.

At this time the planet was repeatedly sheathed in ice in a series of 'snowball Earth' episodes. The greatest of these, the Sturtian glaciation, began 716.5 million years ago. In the space of a few years, land and sea across the globe were swallowed up by sheets of ice that eventually became kilometres thick. They did not melt for another 55 million years. Earth was literally a snowball, like today's Antarctica, from pole to pole.

That, at least, is the story many geologists have come to accept since Joseph Kirschvink of the California Institute of Technology in Pasadena first advanced the idea of snowball Earths in the early 1990s. According to him, ancient glacial deposits laid down at tropical latitudes – for example in north-western Canada, which 700-odd million years ago straddled the equator – tell a story of sea ice between 1.5 and 3 kilometres thick.

The same region also provides clues to the Sturtian glaciation's cause. The Franklin Large Igneous Province, a vast area of volcanic rock covering more than 1 million square kilometres, can be dated to shortly before the glacial layers. It seems the eruption of a supervolcano brought vast volumes of basalt to the surface that quickly weathered under tropical rainstorms – a chemical process that sucked huge amounts of the greenhouse gas carbon dioxide (CO_2) out of the atmosphere (see Chapter 6). Temperatures plunged and the polar ice caps began to advance.

From then on things proceeded with unusual rapidity. As the seas cooled and froze, water vapour, a potent greenhouse gas, evaporated in ever smaller volumes. The big freeze gathered pace towards the equator where temperatures dropped to −50 °C, the sort of cold you can only reliably find today deep in the Antarctic.

Perhaps because of its sheer drama, the snowball Earth idea remains controversial. Some geologists opt for a less harsh 'slushball Earth' variant. But Kirschvink thinks the sheer geographical spread of glacial deposits now dated to the same time tell their own story.

This was not the first time Earth had been shrouded in ice. Another snowball episode had started much earlier, 2.4 billion years ago, when the arrival of microbes that cracked water to fuel photosynthesis sucked vast quantities of CO_2 out of the atmosphere. These episodes highlight just how sensitive the connections are between Earth's climate, geology and biosphere.

Back in the Sturtian, eventually, CO_2 seeping out from undersea volcanoes began to warm things again, and cracks in the ice stayed open. Then, snowball Earth was over almost as quickly as it had begun.

Driest deserts and wettest monsoons

250 million years ago

For extreme vistas and even more extreme weather, Pangaea was the place to be. Earth's most recent supercontinent came together about 300 million years ago and started to break apart 125 million years later. At its height, 250 million years ago, it formed a giant C-shape, with the warm Tethys Sea nestled within the curve. On the opposite side of the planet all that would have been visible was a vast unbroken ocean, Panthalassa.

The centre of the continent close to the equator was a desert. But Earth's greatest desert today has nothing on this one. This point in time is just after the devastating End-Permian mass extinction. One proposed cause is a super-greenhouse climate, which persisted for several million years and rendered much of Pangaea's interior uninhabitable. According to Paul Wignall, who researches palaeoenvironments at the University of Leeds, UK, it would have been normal for the temperature to hit 50 °C.

To the north of this vast swathe of reddish dust stood the mighty Central Pangaean Mountains, players in Pangea's most extreme sight – the mega-monsoon, which dropped along the edge of the Tethys Sea. Monsoon rains happen when moisture-laden sea air is blown on to land and forced upwards, cooling and condensing the water to make rain. The sea was probably warm, about 40 °C, says Wignall. The air, too, was hot, and the hotter air is the more moisture it holds. The scale of the mountains would have forced up huge amounts of warm wet air to great heights, cooling it quickly and unleashing a deluge that would have made today's monsoons look like light showers.

Volcano apocalypse 135 million years ago

Welcome to the furnace, aka the Paraná–Etendeka province, where, if the dinosaurs didn't get you, the volcanism did. The southern remnant of Pangaea, known as Gondwana, had already spent millions of years pulling itself apart, separating what we now know as South America from Africa. That rifting was one of the factors that created this red-hot cataclysm. As the rift worked its way north, Earth's crust became thinner. Meanwhile, a superheated portion of the mantle was welling up, heating the crust from below. Eventually, magma broke through and flooded the land (see Figure 2.4).

The modern-day remnant of this is called the Paraná–Etendeka traps, expanses of basalt covering more than 1.3 million square kilometres of what are today Brazil, Uruguay, Paraguay, Argentina, Namibia and Angola. For the most part, this would be like the volcanism that gave us Iceland – passive, gentle and only rarely explosive. But this period was punctuated by a series of almighty eruptions. The volcanic Explosivity Index has a maximum value of 8, which is described as 'apocalyptic' (one ranking above 'mega-colossal'). It is reserved for any event that ejects more than 1000 cubic kilometres of rock, as the supervolcano Toba did in Indonesia 75,000 years ago. The

FIGURE 2.4 Explosive eruptions were a feature of Gondwana.

Paraná–Etendeka traps produced at least nine apocalyptic eruptions, probably over several million years. They are among the most violent eruptions in Earth's history, as far as we know.

The largest of them spewed at least 8600 cubic kilometres of rock, based on what we can see in South America and Africa today, and perhaps as much as 26,000 cubic kilometres. If you factor in far-flung ash and gases, that's enough material to cover the entire UK to a depth of 100 metres. An event on this scale would incinerate, smother or choke everything for hundreds of kilometres in every direction. Lava from one eruption travelled 650 kilometres.

An almost inconceivable volume of ash would have been lofted into the upper atmosphere, darkening Gondwana's skies for years to come. This ash, combined with sulphate aerosols produced by the eruption, would have reflected solar radiation back into space, quickly plunging the world into volcanic winter, says Sarah Dodd of Imperial College London. The large-scale destruction of plant life would have taken with it the entire regional food chain, wiping out numerous dinosaurs.

Arctic heatwave 55 million years ago

For the previous few million years, Earth had gradually been getting hotter and hotter, and was now on the verge of a planetary heatwave the likes of which have rarely been seen. The Arctic was hot and steamy and home to palm trees. This is the time of the Paleocene–Eocene thermal maximum, or PETM.

Even before the mercury peaked, it was pretty toasty. The poles were essentially ice-free, the deepest reaches of the oceans 8 °C warmer than today, sea levels were roughly 70 metres higher and crocodile-like champsosaurs swam in the Arctic Ocean. The fact that they thrived so close to the North Pole means that water temperatures must have been no less than 5 °C, even in the permanent darkness of winter. (Today's average winter temperatures at the North Pole hover around −34 °C.) Hippopotamus-like *Coryphodon* lurked in the warm swampy forests along the ocean shores.

Fast-forward a few million years and freshwater turtles appeared, which seems bizarre until you consider that the Arctic Basin is almost entirely enclosed by land. River water streaming off the land floated on top of the heavier saltwater, forming what may have been one of the biggest lakes the planet has ever seen. The water would have been a pleasant 23 °C. The other end of the world would also have been experiencing swimsuit weather. At the peak of the PETM ferns appeared on Antarctica, so that's seriously toasty, points out Kate Littler of the University of Exeter in the UK.

All this warmth is the result of a big rise in the concentration of greenhouse gases in the atmosphere, although no one knows what caused it. One possibility is intense volcanic activity, another that deposits of solid methane sitting at the bottom of the sea melted, releasing their load in one great, gassy belch. Or maybe Antarctica's permafrost thawed, releasing a big puff of CO_2. Whatever the cause, after millions of years of gradual warming, temperatures suddenly jumped by at least 5 °C in just 20,000 years.

It was a tough time for life on the sea floor, where an extinction was going on, but life on land seemed to be flourishing. In the lush forests of South-East Asia, a new class of mammal had just evolved, the primates. These looked a bit like tarsiers or bush babies, ate insects, and in the very, very distant future would give rise to the only animal to have occupied all four corners of the planet: humans. Our species is also the only one with the power to trigger something even greater than the PETM: a similar amount of warming but in one-hundredth of the time (see Chapter 10).

Making the Mediterranean 5.33 million years ago

Stand today at Punta de Tarifa, the southernmost point of mainland Europe, and the mountains of Morocco are clearly visible across the Strait of Gibraltar. This busy stretch of water, just 14 kilometres across, is the gateway between the Atlantic Ocean and the Mediterranean Sea, and the closest thing to a border between Africa and mainland Europe.

Yet 5.4 million years ago the picture was very different. The mighty Atlantic was there, but not the Mediterranean. In its place sat a vast basin, glittering with salt crystals and dappled with lakes of hypersaline water. This depression was 2.7 kilometres below sea level at its lowest point. It would have been quite a spectacle: Earth's lowest land today is the Dead Sea basin, a mere 430 metres below sea level.

At the height of the Messinian Salinity Crisis, tectonic movements closed the Strait of Gibraltar, cutting off the Mediterranean. In the hot and dry climate, it took perhaps 1000 years for the sea to evaporate almost completely. Remains from this time can still be found today, under the sea floor and along its shores are thick deposits of salt and gypsum. The basin didn't stay desiccated for long. As time rolled on, the climate grew cooler and wetter, and rivers flowing into the basin turned it into a type of wetland called a lago mare, or 'lake sea'. But, to the west, a cataclysm was brewing.

Some 5.33 million years ago, a combination of tectonic subsidence, erosion and sea-level rise began to reopen the door to the Atlantic. The Zanclean flood – named after the geological age in which it happened – probably started slowly, gradually filling about 10 per cent of the basin over thousands of years. But it ended with a deluge of biblical proportions, according to Daniel Garcia-Castellanos of the Institute of Earth Sciences Jaume Almera in Barcelona, Spain. The rate of inflow suddenly soared, filling the basin completely in the space of a few months, raising the Mediterranean by about 10 metres a day. Every second, a billion cubic metres of water roared past, 5000 times more than the Amazon discharges today. It would have been awe-inspiring.

Astonishingly, this could all happen again. Not enough water flows from rivers into the Mediterranean to compensate for evaporation: it needs the Atlantic to keep it topped up. If tectonic forces were to seal off the strait, the Mediterranean would eventually dry up once more.

Paradise submerged 10,000 years ago

From the Victorian pier at Cromer on the east coast of England, the North Sea looks bleak and uninviting. But nip back

10,000 years and it would have been a very different sight. At the dawn of the Holocene, as the last ice age ended, sea levels were many metres lower than today and Britain was connected to mainland Europe by a fertile plain stretching as far east as Denmark. This area went by the name of Doggerland.

Doggerland has been revealed as a prehistoric paradise of marshes, lakes, rivers – and people. In 2008 University of Bradford archaeologist Vincent Gaffney and colleagues used seismic survey data gathered by a Norwegian oil company to reconstruct this lost world. The result is a map covering 23,000 square kilometres, an area roughly the size of Wales.

At its southern end, Outer Silver Pit Lake, which is now a depression in the floor of the North Sea, filled from the east by the River Thames and from the west by the Rhine. Doggerland's people congregated on its shores to fish, hunt and gather berries. Gaffney describes it as 'prime real estate for hunter-gatherers'. Today, North Sea trawlers occasionally dredge up traces of these people from the seabed – a spear point fashioned from deer bone, for example. But not much else is known about them.

What we do know is that they were victims of climate change. As the world warmed and the glaciers melted, sea levels rose by around 2 metres a century, gradually engulfing low-lying areas. Over a few thousand years, Doggerland transformed into an archipelago. Then, about 8150 years ago a mega-tsunami struck, triggered by a massive undersea landslide off the coast of modern-day Norway, known as the Storegga Slide. A 2014 study estimates that 3000 cubic kilometres of sediment collapsed, probably triggered by an earthquake. It generated a giant tsunami that surged across what was left of Doggerland, according to John Hill of Imperial College London, who led the research.

Any remaining islands of Doggerland would have been inundated, leading Hill and others to suggest that the Storegga Slide sounded the death knell for its people. Ultimately, the result was a cultural separation of Britain from mainland Europe that would last for millennia.

3
Inside and out

Earth is like an onion, with different layers extending from its soils down to its iron heart. Most of these layers are physically inaccessible to us, but that has not stopped us. Ingenuity, experimentation and an ever deeper understanding of nature have enabled us to see the invisible.

Down, down, deeper and down

> 'Descend, bold traveller, into the crater of the jökull of Snæfell, which the shadow of Scartaris touches before the kalends of July, and you will attain the centre of the earth. I did it.'
> Arne Saknussemm

With just this note for guidance, Otto Lidenbrock headed for the Icelandic volcano Snæfellsjökull and thence downwards into the planet. Once below ground he encountered deep oceans, prehistoric creatures, lightning storms and giant insects.

At least, that's what Jules Verne wrote in 1864 in *Journey to the Centre of the Earth*. If only it were that simple to find out about the insides of Earth. So far, we have managed to drill down through only about one-third of the planet's outer crust. Beyond that, we have had to deduce its make-up largely from pressure waves and the way they travel through rocks of different densities.

Seismic waves

Ironically, the most valuable source of these pressure waves has been those most destructive of events – earthquakes. Geologists measure how long it takes shock waves to travel from a quake to various points around the globe. This has revealed that Earth is made up of layers, rather like an onion. On the outside is the thin crust – its depth no more than that of a postage stamp stuck on a football. Underlying that is a mantle, which makes up more than 82 per cent of Earth's volume. Deeper still is the very dense and hot core.

The shock waves from an earthquake spread out in all directions and are reflected or refracted when they meet rock of different densities. If the rock is denser, the pressure waves speed up; if the rock is less dense, they slow down. By determining the path and speed of these seismic waves through Earth, geologists can identify the density and thickness of rocks that lie thousands of kilometres below our feet.

The Irish geophysicist Robert Mallet kick-started the science of seismology in the late nineteenth century. With it, researchers discovered that earthquakes send two main types of waves down into the planet. Primary, or P, waves are similar to sound waves. They alternately compress and expand the medium through which they travel, and can pass through solids, gases and liquids. Secondary, or S, waves oscillate at right angles to their direction of travel. This means they can pass only through solids – liquids and gases have no rigidity to support their sideways motion.

When Andrija Mohorovičić, a Yugoslavian geophysicist, analysed the records of an earthquake in Croatia in 1909, he detected four seismic pulses. Seismographs close to the quake recorded slow-moving S and P waves. In recordings made farther away, these signals soon died out, to be replaced by faster S and P waves.

Mohorovičić interpreted the slow waves as having travelled to the seismometers through the upper layer of the crust. The faster waves, however, must have passed through an underlying layer of denser rocks, which deflected them and increased their velocity. He concluded that a change in density from 2.9 to 3.3 grams per cubic centimetre marked the boundary between Earth's crust and mantle. This boundary, which lies an average of 8 kilometres beneath ocean basins and 32 kilometres below the continents, is now called the Mohorovičić discontinuity or, simply, 'the Moho'.

Shadows inside Earth

As seismologists collected more and more records from earthquakes, they noticed a 'shadow' zone, free of shock waves, between 105 and 142 degrees (see Figure 3.1) from the source

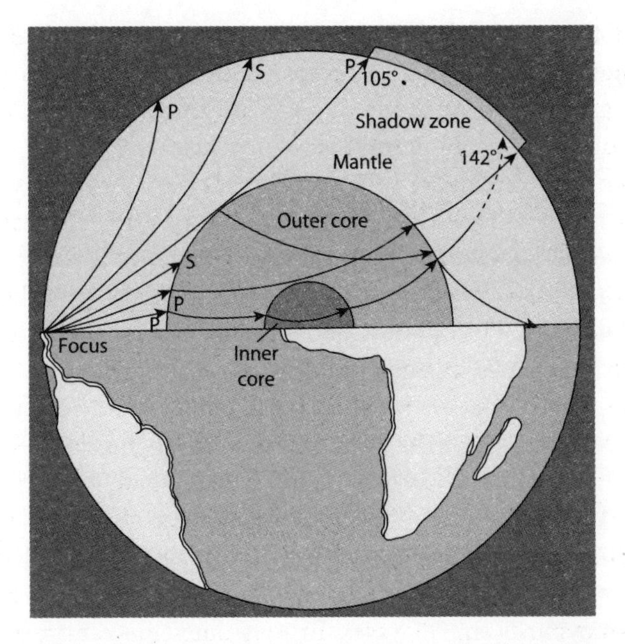

FIGURE 3.1 Seismic waves are stopped dead or refracted by layers of different densities. The liquid outer core casts a shadow starting at 105 degrees from the source of an earthquake.

of the quake. Beyond 142 degrees the waves reappeared. The only way to explain this result was that the waves had passed from a solid into a liquid, stopping the S waves dead and slowing and refracting the P waves. The density change from 5.5 to 10 grams per cubic centimetre at a depth of 2900 kilometres was designated as the boundary between mantle and core.

Only later were much fainter waves detected in the shadow zone. In 1936 the Dutch seismologist Inge Lehmann suggested that a further change in density occurred at about 2200 kilometres into the core. This change would accelerate the P waves and bend some of them so that they appeared in the shadow region. She concluded that inside Earth there is an inner core of a very dense solid. We estimate that the density changes from 12.3 to 13.3 grams per cubic centimetre.

Extreme conditions

The picture of Earth we now have is of a series of concentric layers which become progressively denser towards the centre (see Figure 3.2). Two opposing factors control this density.

The first is temperature, which works to soften and or melt rocks. Much of the inner Earth is hot from the energy produced by the decay of radioactive elements in the rocks. At the centre of Earth the temperature is perhaps 3000 °C, falling to 375 °C at the mantle–crust boundary. The second factor is pressure, which tends to solidify rocks. The deeper you go, the greater the weight of overlying rock and the higher the pressure.

In effect, close to the cold surface, rocks are mainly solid and brittle. This lithosphere, which takes in the crust and upper mantle, reaches down to about 60 kilometres. At this point, seismic waves slow down, indicating a drop in density. This is the 'strength-lacking' asthenosphere, where heat from radioactivity cannot easily dissipate and the rocks tend to melt. Its toffee-like consistency extends down to almost 200 kilometres.

Below the asthenosphere, the seismic waves at first accelerate rapidly and then more slowly for about 2100 kilometres. This is the 'mesosphere', where pressure conquers the increasing heat to make the rocks more rigid so they can only 'creep' very slowly. At the mantle–core boundary, where the S waves peter out, the temperature is at first high enough to counteract the immense pressure and, for some 2200 kilometres, the outer core is liquid. But at the centre, pressure again prevails to create a solid inner core with a radius of 1270 kilometres.

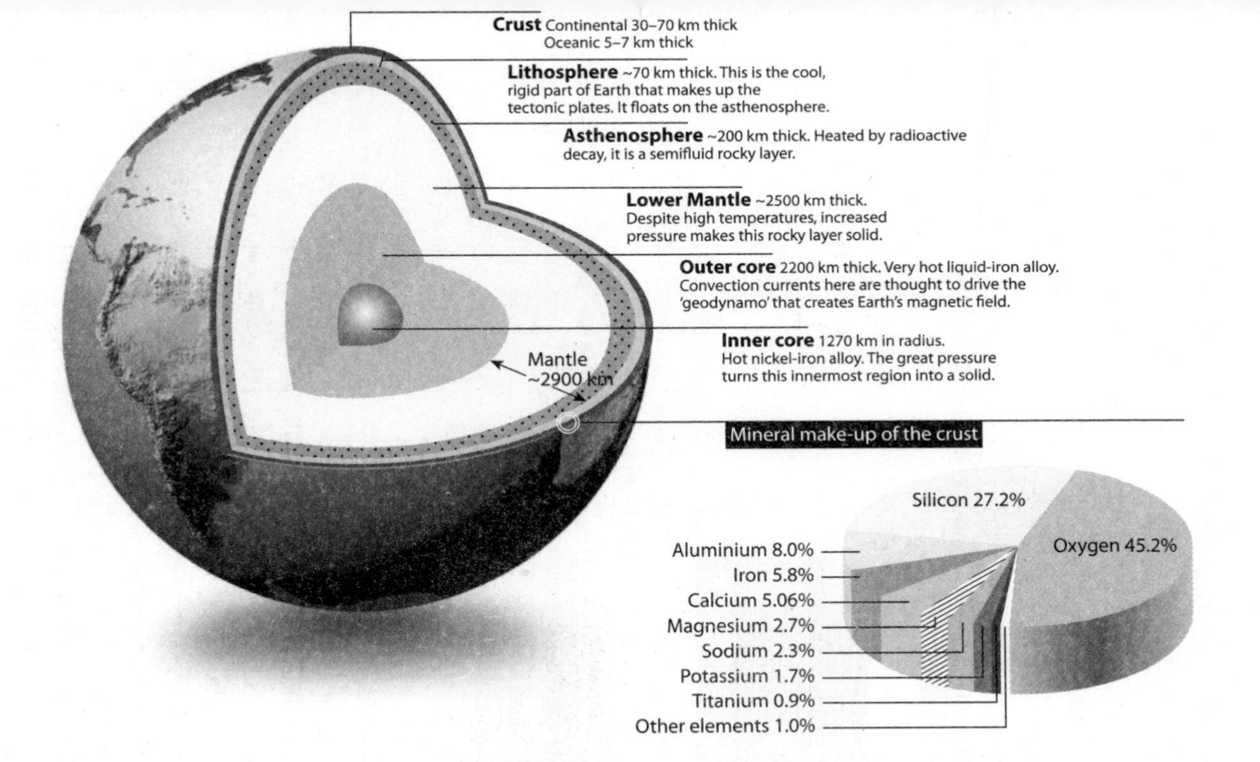

FIGURE 3.2 Seismic surveys have given us a good idea of the layered structure of Earth. The make-up of the crust is a job for chemical analysis.

Mysterious mantle

It's one thing to know about the physical state of Earth's innards. It's another to know what it's made of. The crust, we can analyse. In continental areas we find silicon and aluminium in abundance; combined with oxygen these minerals make up the most common rock, granite. Below the oceans, and underlying continental granite, we find basalt rock in which silicon, iron and magnesium predominate.

There our certainty ends. The mantle, which makes up about two-thirds of Earth's mass, is still terra incognita. We have no pristine samples. Some rocks from below the crust have reached the surface but they are all contaminated. For example, rare rocks called mantle nodules have erupted in volcanoes, showing that the mantle is made of minerals including olivine and pyroxene, which form only at high pressure and contain little silicate but lots of magnesium and iron.

And, in some parts of the ocean floor, rocks that were once part of the mantle lie exposed, but contact with seawater has changed their composition dramatically. Without fresh samples, geologists struggle to confirm even simple facts about what exactly the mantle is made of, how it formed and how it works.

Instead, they have had to piece together their theories about the mantle using indirect evidence and knowledge gleaned from lab experiments. These show, for example, that while silicon, magnesium and iron exist in minerals such as olivine and pyroxene in the upper mantle, at pressures found deeper down, the atoms would rearrange themselves into more compact 'high-pressure minerals', altering the composition of rock. In the lower mantle, the minerals are likely to break down into simple oxides.

Further clues to composition have come from meteorites, which were forged from the same cosmic debris as our planet. Stony meteorites are likely to represent mantle-like material and iron meteorites the core. These metallic projectiles contain mostly iron, iron sulphide, nickel, platinum, and traces of iridium.

Messengers from the underworld

Discovering the composition of the mantle would improve immeasurably our knowledge of Earth's chemical inventory and give us clues about conditions when it formed. One way to interrogate the mantle is to use the neutral, near-massless particles called neutrinos. They – or more precisely an antimatter variant called electron antineutrinos – are spewed out in vast numbers by the radioactive decay of uranium, thorium and other radioisotopes in rocks deep in Earth's interior.

This is helpful because like iron, silicon and all those other elements, uranium and thorium were present, albeit in smaller amounts, in the solar nebula from which Earth formed, and would have condensed out in different amounts at different temperatures. If we knew how much uranium and thorium went into making Earth, we would know what the primaeval conditions were like and be able to estimate how much of everything else we would find inside. By tracing where in the mantle uranium and thorium are distributed – whether homogeneously, in patches of different compositions, or in layers – we can also learn about our planet's inner dynamics.

There is no better way of doing all this than by counting the 'geoneutrinos' that different radioactive isotopes produce. Because they hardly interact with normal matter, these particles race unimpeded through Earth's interior, allowing detectors near the surface to snag them as they leave, in principle, at least. In practice, that same flightiness makes neutrinos far more likely to pass through our detectors too. Geoneutrino hunting takes skill and a lot of patience.

Fortunately, we have spent more than a decade developing these detectors. The Kamioka Liquid-Scintillator Antineutrino Detector (KamLAND), which came into service near the central Japanese city of Hida in 2002, consists of 1000 tonnes of a transparent liquid which, when hit by a neutrino, emits a flash of light. It is situated 1 kilometre down, the better to shield it from cosmic-ray muons, whose signals mimic those of neutrinos (see Figure 3.3).

In 2005 KamLAND saw the first, faint signal of electron antineutrinos from Earth's bowels, but it was drowned in a din of antineutrinos produced by nearby nuclear power plants. In 2007 a detector upgrade and the temporary shutdown of one of the largest nuclear plants allowed the planet's signal to shine through. By the end of 2009 KamLAND had recorded 106 electron antineutrinos with the right energy to have come from decays of uranium and thorium within Earth.

FIGURE 3.3 KamLAND has an 18-metre vessel lined with light detectors. It is filled with a liquid 'scintillator', which sends a flash of light when an antineutrino strikes.

The Borexino experiment has also been getting glimpses. Situated at the Gran Sasso National Laboratory in central Italy, this smaller detector was built to pick up neutrinos from nuclear processes in the sun. Combining data from the two experiments was enough to produce the first concrete geophysical predictions from geoneutrinos alone: that the decay of uranium and thorium in the mantle and crust contributes about 20 terawatts to the heat escaping from Earth's interior.

These are the sorts of numbers we need if we are to start outlining what lies beneath. Earth radiates about 46 terawatts of heat through its surface, from two sources: 'radiogenic' heat produced in radioactive decays, and 'primordial' heat stored up during Earth's formation as particles collided and iron sank to the core. Establishing how much surface heat comes from each source has wide ramifications for our picture of Earth.

For example, if material in the mantle is convecting slowly, or in layers with limited heat transfer between them, little primordial heat will be transported from Earth's innards to its surface. If so, the lion's share of Earth's heat flux – 30 terawatts or more – must be of radiogenic origin. The neutrino experiments suggest that the true figure is lower, implying that the mantle is mixing relatively thoroughly.

Mineral puzzles

The radiogenic heat flux also indicates that the planet has an overall uranium content of some 20 parts per billion. Exposed mantle rocks contain similar amounts of uranium, suggesting that they are indeed representative of the mantle, and backing up the idea that the entire mantle is mixing efficiently. But it also hides a puzzle. The exposed mantle rocks are dominated by the magnesium iron silicate mineral olivine, and their uranium content is appreciably higher than that of a class of stony meteorite called enstatite chondrites. These meteorites have long been thought to be representative of the material that made Earth, yet they are dominated by another silicate material, pyroxene. That raises the question of where on Earth this pyroxene-dominated material is – hidden in pockets deep in the mantle, perhaps? Or is the planet's composition different from that of enstatite chondrites?

The ratio of olivine to pyroxene in Earth's mantle is crucial to pinning down where and when the planet formed in the solar nebula. Olivine would have precipitated out at a slightly higher temperature than pyroxene, so there would have been more of it closer to the Sun, or earlier in the planetary construction process when temperatures were higher.

We are still a way away from the answers. With the numbers of geoneutrinos spotted so far, there is a lot of wiggle room in the estimate of radiogenic heat flux: the 20 terawatt figure comes with a quoted error of about ±9 terawatts, making it hard to discount any scenario of mantle composition or mixing. KamLAND and Borexino alone are unlikely to put the debate to rest.

A third detector, due to switch on in 2018, could make a decisive difference. This is SNO+, situated deep underground at the Sudbury Neutrino Observatory in Ontario, Canada. It is about the same size as KamLAND but, because it is under 2 kilometres of rock, will be better protected from cosmic-ray muons. And it is not surrounded by nuclear reactors. With lower background counts, SNO+ should observe geoneutrinos by the bucketful – by neutrino standards, anyway.

That is just the beginning. Ideally, we want to map where geoneutrinos originate, and so get a finer-grained picture of the distribution of uranium and thorium and the homogeneity and mixing of the mantle. That means screening out geoneutrinos from other sources, such as the crust and core, which will take a network of detectors looking for neutrinos coming up from different places and at different angles. This would allow us to find out more about peculiar regions of the mantle, such as 'superplumes' below Africa and the Pacific Ocean that have been invoked to explain anomalous areas of volcanism (see Chapter 5).

The velocity of seismic waves drops dramatically through these superplumes, suggesting that they are less viscous and so perhaps hotter. That might be because they contain larger amounts of decaying uranium and thorium. If so, they should be geoneutrino hotspots.

Go shallow, go deep

It's not only cosmic rays, the Sun and nuclear reactions that can confound geoneutrino signals. To map goings-on inside Earth's mantle, we also need to rule out neutrinos from Earth's crust and core. The crust is thin compared to the mantle, but its proximity to underground detectors means its geoneutrino signal can overwhelm the one from the mantle.

To reduce the impact of this noise, in Canada researchers set out to characterize the crust's rock formations right down to the mantle boundary around the SNO+ neutrino experiment. The aim was to estimate how much uranium and thorium is there, and so how many neutrinos their decays are likely to produce.

What about neutrinos from the core? Do they generate noise, too? Not too long ago, geophysicists thought it likely that there was enough uranium in the core to make it a giant nuclear fission reactor. But simulations show that at the high temperatures and pressures found in the magma oceans that filled early Earth, uranium almost exclusively prefers the company of elements found in mantle-like rocks to the iron and nickel of the core.

Heart of the planet

Seismic wave measurements, computer models and lab experiments that mimic the core's extreme conditions all provide a reasonable picture of how things work in Earth's deepest reaches.

From seismic surveys, we know that the core starts 2890 kilometres down, and that its radius is 3470 kilometres. It is composed of two layers, the molten iron outer core and the solid inner core, which is made of nickel and iron and is roughly the size of the Moon.

It hasn't always been this way. Initially, the planet was just one big jumble with no obvious structure. Then the heaviest elements, mostly iron and a little nickel, settled towards the centre. Exactly when and how this happened is still up for debate. One idea is that the core formed suddenly, in an avalanche towards the centre. Others believe that the iron slowly trickled down.

Radioactive isotopes measured in volcanic rocks that originated deep in Earth indicate that the core formed when the planet was somewhere between 30 and 100 million years old. By 3.5 billion years ago, swirling motion in the liquid iron core had set up a magnetic field. Then, between 1.5 and 1 billion years ago, the centre of the core cooled enough to crystallize into a solid inner core.

One mystery surrounding the core has been solved. It has been known for some time that seismic waves travel faster through the eastern side of the core than the western, but nobody could work out why. Now simulations have shown that this is most likely due to swirling eddies of liquid iron in the outer core that pull down cooler material from near the core–mantle boundary and plaster it on to the solid inner core. For the past 300 million years most of the iron eddies have been under Asia, causing the inner core to grow to around 100 kilometres larger on its eastern side than on the west.

This could have implications for Earth's magnetic field, which most researchers think is generated by convection in the liquid outer core. Some think that turbulence caused by the growing inner core may, over time, make the magnetic field less stable and more likely to flip, causing Earth's north and south magnetic poles to swap places (see Chapter 4).

When that happens – as it has done in the past – the planet is left temporarily unprotected from the torrent of energetic particles streaming out from the Sun, known as the solar wind. The result of that could be disastrous – at best it might fry all our electronic equipment, at worst it could strip off our life-supporting atmosphere. The good news, or bad, depending on your point of view, is that nobody knows when that flip will happen.

Invisible shield

Earth's magnetic field is thought to be generated by the most basic of physical processes. Electrons flowing through the fluid core generate an electric current, which in turn creates a magnetic field. In other words, the core is a giant, self-sustaining dynamo.

Simple it may be, but there is a problem with it. In recent years, evidence has mounted that the dynamo must be a new phenomenon, comparatively speaking. Yet magnetism in ancient rocks clearly shows that a field has existed for most of Earth's history. Reconciling this apparent paradox is forcing a rethink of just what happens inside our planet.

As we've just seen, received wisdom says that as the early Earth cooled its dense iron sank slowly to the centre where high temperatures melted it. Next, thermal convection – the process by which hot fluid rises and cool, dense liquid sinks – kicked in. This motion set the dynamo into action.

Then came a complication. Earth cooled sufficiently to allow some of the core's molten iron to solidify. At the extreme temperatures and pressures inside the planet, the core began to freeze from the inside out. According to most estimates, this process started after 1.5 billion years ago. Today, the inner core is a solid iron ball more than 1200 kilometres in radius and is growing as Earth cools.

Fortunately, the freezing of the core kick-started another effect that kept the magnetic dynamo humming. As the inner core grows, it expels any lighter elements. The same thing happens when salt water freezes: salt doesn't fit into ice's crystal structure, so it gets pushed out, leaving freshwater ice floating on extra-salty water.

Similarly, inside Earth, the solid inner core is almost pure iron, while the surrounding liquid contains iron plus nickel and a smattering of sulfur, oxygen and other lighter elements. These extra ingredients make the liquid surrounding the inner core less dense, so it rises. Away from the inner core, heavier iron-rich material sinks, setting the outer core churning in a process called compositional convection.

The mainstream view, then, is that thermal and compositional convection have kept the dynamo turning for most of Earth's history. But here's where the trouble starts. In the past few years, researchers have started to doubt whether thermal convection ever happened – or, if it did, whether it would have been strong enough to power the planet's magnetic field.

The problem lies in the way heat travels. Convection requires a temperature difference: in a pot of boiling water, the bottom is hotter than the top. This can only happen because water is a poor conductor of heat. Good conductors quickly equalize temperatures and stop convection dead. And there's the rub:

FIGURE 3.4 Polar auroras are the most obvious and beautiful evidence of the existence of Earth's magnetic field.

growing evidence suggests that the core is a better conductor than we originally thought.

In 2012 two separate computer models predicted that the liquid core must be twice as conductive as previously believed. Then, in 2016, Kei Hirose's group at the Tokyo Institute of Technology in Japan measured the thermal conductivity of iron under pressures comparable to those in the core. The results matched the two predictions, suggesting that Earth's magnetic field could only have emerged with the first solidification of the core after 1.5 billion years ago.

Except it can't have done. We know Earth's magnetic field existed long before then because its presence is recorded in ancient volcanic rocks: when molten rock solidifies, its magnetic minerals align with Earth's field (see Chapter 4). Solid geological evidence has found these prehistoric compasses in rocks at least 3.45 billion years old, leaving a substantial gap when neither thermal nor compositional convection could have sustained the dynamo. So what did?

The third way

David Stevenson at the California Institute of Technology and others think they have a solution, a third form of convection that doesn't rely on activity around the inner core. Instead, they focus on events at the outer boundary of the liquid core. As this cools, lighter elements dissolved in the iron would precipitate out and be absorbed into the mantle. The denser liquid left behind would then sink, triggering convection.

Work is now under way to identify the main driver of this process, and the prime suspects are magnesium or silicon. Magnesium doesn't readily dissolve in iron and so would precipitate out easily. But for it to be dissolved in the first place would have required very intense heat, presumably produced by the ferocious collisions around the time of Earth's creation.

Silicon is even more abundant and so would be likely to dominate deep in the planet. Experiments by Hirose's group show that silicon dioxide crystallizes easily in the core, without the need for any external processes. He thinks silicon dioxide is the most likely driver of this new form of convection.

Some researchers have even suggested that convection may not drive the dynamo at all. Instead, Earth's wobbling rotation could jostle the molten iron; or the Moon's gravity could tug the liquid core in the same way that it causes ocean tides. But these ideas are not considered mainstream.

It seems more likely that a new form of convection is at work deep in Earth. Although whether that alone can resolve the magnetic paradox is still unresolved. Fortunately, we shouldn't have to wait long for the answer – this is a fast-moving field.

The loose, soft cloak of life

Our deep dive into the planet has overlooked a layer that cannot go without mention. It's a layer most of us see every day but, unless we are farmers or gardeners, probably take for granted. Yet, without the pedosphere – the soil, that is – most plants would be unable to grow and the food chain that supports most animals, including us, would be severed at its base. In short, life on land would be a mere shadow of what it is today.

From a planetary perspective, soil is a vast interface zone that covers much of Earth's land surface where the atmosphere, the hydrosphere, the biosphere and the geosphere all meet. It is both a fascinating and an important environment, and a very complex one. Not only does soil reside at the point of overlap between these four spheres but it owes its very existence to interactions between them.

Soil is made up of solids which are in part the products of the weathering of rocks and in part the products of biological activity – the decay of plant and animal debris. Soil is highly porous, with typically a 50:50 mix of solids and pore spaces. The pore spaces contain variable amounts of water and air, depending on how wet or dry the soil is.

Soils are formed by a combination of processes. Physical weathering, by repeated freezing and thawing and the work of wind, water and ice, breaks the bedrock into particles, for example. Chemical weathering alters the make-up of the minerals from which rocks are made. Soil also arises from the decomposition of plant and animal matter; and through the movement of solids and dissolved chemicals by water.

One of the key functions of soils is life support, and soil itself is teeming with life, mostly microorganisms which cannot be seen with the naked eye, but more obviously the larger animals, such as earthworms, and the plant life that live in and on it. All this biological activity results from the ability of the soil to provide the necessities of life: shelter, food and water.

For plants, and the many animals that consume them, nutrition depends on the intake of a number of vital elements, such as nitrogen, phosphorus, potassium and calcium, which they require in varying amounts. Processes such as rock and mineral weathering and the breakdown of organic materials release these elements in soluble forms into the soil water. They become available for use by soil organisms and by plants, which absorb them through their roots.

Endangered resource

The processes that take place within soils are influenced by environmental factors, such as the nature of the underlying geology, the local vegetation, the climate and the topography. This means that in different places the processes combine in different ways to produce a variety of soils.

Soils are classified by the features they exhibit when viewed in vertical sections known as soil profiles. In temperate zones, for example, sandy materials derived from sandstone, under coniferous forests on steep slopes, may develop into podzols, which have a characteristic banded structure created by the loss and redistribution of material by percolating water. In tropical climates, strongly weathered basaltic rocks under rainforest may generate deep, red, iron-oxide-rich oxisols.

The classification system for soils is every bit as elaborate as the one we use to categorize life forms. In the USA alone, more

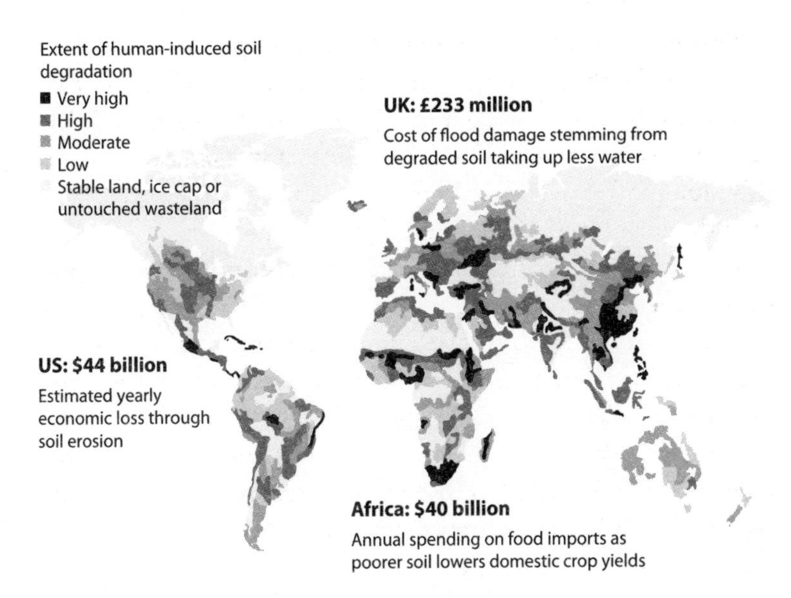

FIGURE 3.5 Every year the world loses an area of healthy soil the size of Louisiana. This has major economic consequences.

than 20,000 soil types have been catalogued. Many are facing extinction. Indeed, estimates suggest that more than a third of the world's top layer is endangered. And that is not the full extent of the damage. Most of the world's soils are in only fair, poor or very poor condition, according to the UN's 2015 report *Status of the World's Soil Resources*. Erosion carries away up to 40 billion tonnes of topsoil every year, nutrients in soil are being depleted and human-induced salinity has spread to an estimated 760,000 square kilometres – the size of all the arable land in Brazil.

If we don't slow the decline, all farmable soil could be gone in 60 years. Given that soil grows 95 per cent of our food, and sustains human life in several other surprising ways, this is a huge problem. Many would argue that soil degradation is the most critical environmental threat to humans.

Explosive extinction

Abruptic durixeralfs is a subgroup of soil that nestles between prime agricultural land in the western USA. It is endangered, partly because it is of little direct use for farming: it has an unfortunate tendency to form a hardpan, a dense, compacted layer that repels both roots and water. Frustrated farmers have resorted to using explosives to blow it out of the way – a dramatic method of soil extinction, but only one of many.

Services from soil

The degradation of the world's soils has been a disaster in slow motion which scientists have been warning about for decades. All the while, our understanding of just how crucial soil is has only grown. Soil does not only grow our food. A single gram might contain 100 million bacteria, 10 million viruses and 1000 fungi. This microscopic menagerie is the source of nearly all of our antibiotics, and could be our best hope in the fight against antibiotic-resistant bacteria. Well-conditioned soil can, for example, suppress the growth of pathogenic bacteria.

Soil is also a surprising ally against climate change. When nematodes and microorganisms within soil digest dead animals and plants, they lock in their carbon content. Even in their degraded state, it is estimated that the world's soils hold three times the amount of carbon as the entire atmosphere. Water filtering and storage are other talents lost when soils degrade. A UK government report from 2012 suggests that soil degradation costs the country £233 million in flood damage every year.

The greatest threat

It is small wonder, then, that endangered soil is making ecologists so nervous. Soil's biggest problem by far is agriculture. In the wild, nutrients removed by plants are returned to the soil when they die and decay to form rich humus. Humans tend not to return unused parts of harvested crops to replenish those nutrients.

We realized this long ago and developed strategies such as leaving fields fallow, or rotating crops that needed different nutrients, thereby keeping the soil in balance. Growing peas and beans can even add nitrogen to the ground: nodules in their roots host rhizobia bacteria, which grab atmospheric nitrogen and convert it into absorbable nitrates.

But these practices became inconvenient as populations grew and agriculture was mechanized. A solution came in the early twentieth century with the Haber–Bosch process for manufacturing ammonium nitrate, a rich source of available nitrogen. Farmers have been plying their fields with this synthetic fertilizer ever since.

But over the past few decades it has become clear that this wasn't such a bright idea. Chemical fertilizers release polluting nitrous oxide into the atmosphere and any surplus is often washed away, transporting nitrogen into rivers, where it can fuel damaging algal blooms. More recently, we have found that indiscriminate fertilizer use hurts the soil itself, turning it acidic and salty. It also suppresses symbiotic relationships between fungi and plant roots, and can even turn beneficial bacteria against each other.

So, in many ways, fertilizers speed towards extinction the very soil they are supposed to nourish. They are good for growing plants. Yet they hide the nature and extent of soil degradation. Worse still, when soil stops being productive, famers tend to add more fertilizer.

Solutions to soil degradation

A wholesale return to leaving fields fallow is unlikely to happen, so what else can be done to help save our soils?

One possible solution is to make fertilizers smarter. Plants effectively 'tell' microbes when they need nitrogen, and the microbes respond by freeing nitrogen from organic matter. In 2011 Carlos Monreal of Carleton University in Ottawa, Canada, and his colleagues identified five compounds that plants exude when they need nitrogen.

The researchers created a fertilizer that keeps its payload locked up until it encounters these chemical signals. The key was aptamers, short strands of DNA that bind to specific chemicals, in much the same way as antibodies do. They built a scaffold containing the aptamers around a tiny parcel of fertilizer. In the presence of one of the plant-signalling compounds, the aptamers bind to it, break the scaffold and release its contents.

Another school of thought suggests that we should dispense entirely with artificial fertilizers and encourage the soil's own microbial resources: its microbiome. Pius Floris is a pioneer in this field. Running a tree-care business in the Netherlands, he realized that the best way to ensure that his trees flourished was to take care of the soil. So he developed a 'universal recipe' of beneficial bacteria, mycorrhizal fungi and humus that adheres to plant roots and helps them to extract nutrients.

In tests, when Floris's mix was added to desert-like test plots, they began to sprout plants with healthy foliage and roots. The few plants that grew in control plots, dosed with traditional pesticides and fertilizers, were small and stunted.

Innovative as they are, these measures will at best make a small dent in the global soil degradation problem. Broader action is needed. But, before that can happen, scientists need ways of presenting the issues to governments and the public. Pamela Chasek at the International Institute for Sustainable Development, a think tank based in Winnipeg, Canada, and her colleagues have proposed a user-friendly goal of 'zero net land degradation'. Like the idea of carbon neutrality, it is an easily understood target that could help shape expectations and encourage action. It would offer a banner under which projects like Monreal's and Floris's could rally.

4
Plates, quakes and eruptions

Plate tectonics is one of those ideas that seems so simple as to be obvious. So it may be surprising to realize that it was accepted only in the 1960s. From that point on, it has transformed geology in a way not unlike how Darwin's evolution changed biology. Everything suddenly made sense.

A groundbreaking idea

Earth is filled with a viscous mantle on which floats a handful of rigid plates that are constantly rubbing up against each other. That is the essence of plate tectonics.

It is such a simple idea, yet its implications are many. The theory tells us about Earth's history, why oceans open and close, mountains rise and fall and continents gather and split. It reveals where volcanoes and earthquakes appear, and why. Plate movements created many of the oil, gas and mineral deposits that we find on Earth, after being squashed and baked to just the right degree.

Not insignificantly, plate interactions recycle water and carbon, and keep our climate constant, creating an environment that has mostly been good for life. The theory of plate tectonics turned geology from a science of collecting, classifying and cataloguing into one of processes, where predictions could be made and tested.

A tough jigsaw to piece together

In the seventeenth century the English philosopher Francis Bacon noted that the outline of the eastern side of the Americas and the western edge of Africa seemed to fit together like pieces of a giant jigsaw.

In the centuries that followed, settlers in the New World discovered huge deposits of coal in the American continent whose position seemed to match deposits in Europe. Scientists found the fossilized remains of identical species of plants and animals on both sides of the Atlantic. Gradually the notion emerged that the continents might once have been one and were drifting apart.

In 1912 the German geophysicist Alfred Wegener gave this idea a name: 'continental drift'. But he lacked a mechanism to make the continents move, and the idea was at first ridiculed.

Then, in 1928, a professor of geology at Durham University, UK, Arthur Holmes, suggested that there might be convection currents in the upper mantle. The American geologist Harry Hess extended this idea to create the concept of sea-floor spreading. His idea was that convection would force molten basalt, or magma, to well up and open long cracks in the overlying ocean crust. As the magma flowed from these gashes it would cool and spread, creating the great ridges that run beneath the world's oceans.

Scientists were sceptical and relented only in the 1960s, following magnetic surveys of the Mid-Atlantic Ridge. These showed that rocks on the ocean floor were magnetized in alternate directions in a series of bands parallel to the ridge. The pattern of bands appeared identical on both sides of the ridge.

Researchers explained this pattern by arguing that, as magma solidified on the ocean floor, its minerals became magnetized in the direction of Earth's prevailing magnetic field. A later injection of magma would then split the top strip of solidified basalt lengthways in two. If Earth's magnetic field had reversed in the meantime, then the new strip of basalt would be magnetized in the opposite direction to its predecessor.

This explanation – plus the increasing age of rocks at increasing distances from the ridge – showed that as the mid-ocean ridges continually added material to the ocean floors, continents once joined could become separated by huge oceans.

In 1965 the Canadian geophysicist John Tuzo Wilson brought together continental drift and sea-floor spreading into a single concept of mobile belts and rigid plates. In 1967 American geophysicists added another concept – underthrusting, or 'subduction' as it's now known – where one block of crust is drawn beneath another (see Chapter 1).

From these ideas grew the grand theory of plate tectonics: that Earth's outer surface, the lithosphere, consists of seven

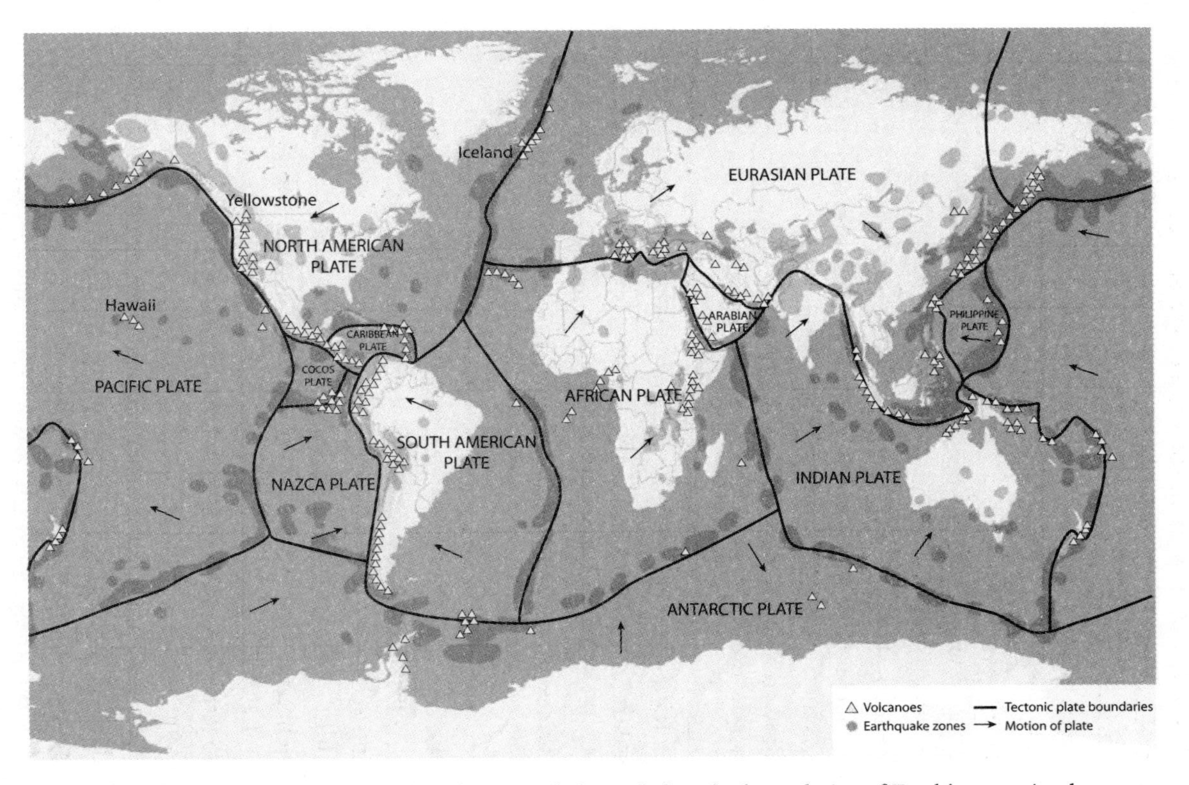

FIGURE 4.1 Volcanoes and earthquake zones help to define the boundaries of Earth's tectonic plates.

large plates, and several smaller ones, which move over the hot, partially molten asthenosphere. As they do so, they carry the oceans and continents with them.

Making plates predictable

In 1967, while hippies gathered in San Francisco to celebrate the counterculture, a young geologist was working in the south of California on an idea that would revolutionize Earth sciences. Dan McKenzie spent the Summer of Love figuring out the maths behind plate tectonics.

McKenzie, who is now at the University of Cambridge, had a moment of insight: he realized that the plates were rigid. It doesn't seem much, but by treating plates as rigid it became possible to think of them as geometric figures, like paving stones on a sphere. He and his colleague Bob Parker saw that the relative motion between two plates could be treated as though it were a rotation about a fixed point – an idea that goes back to the eighteenth-century Swiss mathematician Leonhard Euler.

This enabled the pair to calculate the relative motion of plates in the North Pacific. They found it tallied exactly with earthquake activity in the region, showing that their calculations were correct and that this movement must not only be happening but also be the likely cause of the earthquakes.

It was the last piece of the puzzle. Suddenly, they could see not just that plates shifted, but also that each was moving relative to others. The entire surface of Earth was covered in rigid plates jammed up against each other, yet still in motion. McKenzie and Parker knew they had hit on something big.

As it happened, they weren't the only ones. Unknown to McKenzie, Jason Morgan at Princeton University had arrived at exactly the same conclusion and had given a talk on the subject early in 1967. Morgan could even identify three types of plate boundary: the ridges where new crust is formed; trenches where crust disappears, and faults where crust neither disappears nor appears.

Historians accord ownership of the theory to both McKenzie and Morgan, who is now a visiting scholar at Harvard University. But McKenzie says Morgan has priority. 'Morgan talked about it before I had even thought about it,' he says.

FIGURE 4.2 Dan McKenzie straddles the San Andreas Fault in 1967.

Where all the action happens

Each plate is rigid, deforming only at its edges. At these boundaries, the plates diverge, converge or slide past one another, but very little changes in the middle of the plates. In fact, some of Earth's most dynamic features, such as earthquakes and volcanoes, define the plate boundaries.

At mid-ocean ridges, the plates diverge as new ocean crust is laid down. They resemble conveyor belts always moving the ridges towards ocean shores. The ridges themselves are like continuous elongated volcanoes. Along their lengths, shallow earthquakes at a depth of less than 50 kilometres are common.

Zones where two plates run past each other are prone to shallow earthquakes, sometimes of high magnitude. California's San Andreas Fault, where the American plate is moving south against the northwards-moving Pacific plate, is a case in point.

The most dramatic events take place where two plates converge – that is, collide head on. Here, rocks crumple and mountains are thrown up. A deep trench forms and one plate – almost exclusively the denser ocean crust – is drawn downwards, subducted, into the asthenosphere. So ocean crust, created at a mid-ocean ridge, is consumed back into Earth, and rarely lasts as long as 400 million years (see Chapter 5).

The ocean crust is cool when it starts to descend into the hotter mantle. As it moves and bends, it deforms in a series of short sharp shocks, generating earthquakes that can occur at depths of 700 kilometres. These quakes continue until the descending slab warms up enough to flow rather than crack.

Not all the material that dives into a subduction zone goes all the way down. The edge of the continent can act like the cutting blade of a plane, shaving great masses of sediment from the top of the ocean plate as it descends. Some of these shavings of sediment build up to form islands: parts of Japan began in this way.

Sediments that do descend are among the first parts of the slab to respond to the increasing temperature. Volatile constituents such as water and CO_2 rise upwards into the mantle where they change its composition enough to make it melt. The resulting magma moves upwards and may eventually erupt from volcanoes. Japan's volcanoes are the result of this mechanism.

Know your volcano

Subduction zone volcanoes form a line parallel to plate boundaries, above the descending slab where the mantle melts. Most of the volcanoes of the 'ring of fire', which encircles the Pacific Ocean, are created by subduction, beneath either oceanic or continental crust. Subduction in the oceans produces chains of volcanic islands known as island arcs. The Aleutian Islands, a chain of islands extending southwards from the Alaska Peninsula, are a classic example of an island arc (see Figure 4.3).

The exceptions to the rule that volcanoes cluster at plate boundaries are isolated chains of volcanic islands in the middle of the oceans. Hawaii is just one in a line of increasingly older volcanic islands stretching north-west across the Pacific Ocean. The chain continues as a series of even older volcanoes, which are now under water, the Emperor Seamount Chain.

These volcanoes are thought to arise when a plate moves over a plume of hot magma rising from deep in the mantle. Imagine a piece of paper sliding above a candle: the volcanoes are the equivalent of scorch marks on the paper (see Chapter 5). These island chains provide important clues about the speed and direction of a moving plate. Hot spots beneath continents can also produce isolated volcanoes.

Different volcanoes erupt in different styles. Some, such as Etna in Sicily, mostly smoke and steam, and only occasionally

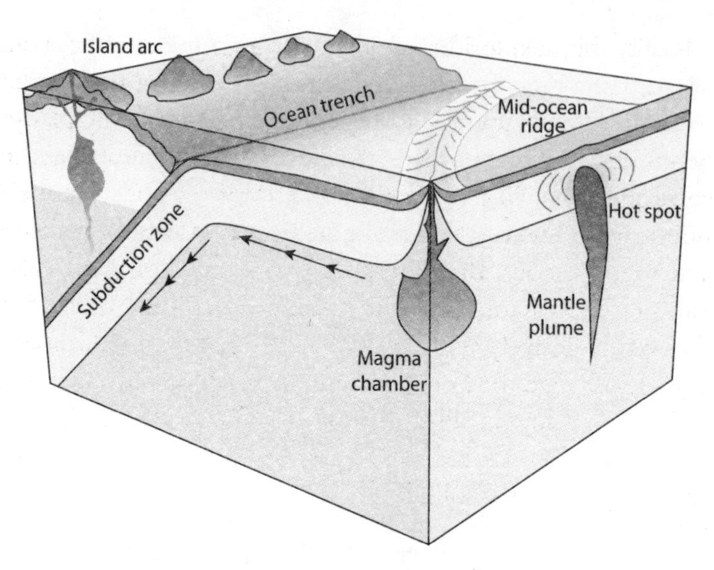

FIGURE 4.3 Oceanic crust forms at mid-ocean ridges and is destroyed at subduction zones. If it is pulled beneath another piece of oceanic crust, the result is an arc, made up of volcanic islands.

disgorge streams and showers of pulverized rock. Hawaiian volcanoes often produce lava lakes and Iceland has fountains of fire that spurt many metres into the air.

Explosive eruptions blow rock, both solid and molten, into ash – tiny fragments of rock and glass – and sometimes coarser debris – rock fragments and pumice – collectively called pyroclastics. Big bangs happen for a variety of reasons. Water infiltrating hot rocks is one cause: water vaporizes rapidly, increasing pressure until the surrounding rock explodes. Other gases can do the same. These dissolve easily in molten rock deep underground, where the pressure is high. As the magma rises to the surface, the pressure drops and gases start to become insoluble, forming bubbles, in the same way that bubbles appear in a bottle of fizzy drink when you release the top.

Really big explosions tend to happen where the rising magma has a high viscosity. Viscous magma traps the emerging gas until it builds up to very high pressures, eventually blasting the molten rock to smithereens. Rocks rich in silicates are the worst offenders. Silica minerals bond to form molecular chains and sheets and when these rocks are heated the bonds make the melt more viscous. Broadly speaking, subduction zone volcanoes produce viscous lava with a high silica content and tend to erupt explosively with a lot of ash. In contrast, volcanoes at mid-ocean ridges or hot spots tend to produce relatively fluid, basaltic lava – low in silica – as in Iceland and Hawaii. In these 'hot, runny' magmas, bubbles of gas rise to the surface or freeze within the rock as it solidifies.

Can we predict volcanic eruptions?

We are getting much better at predicting when volcanoes will erupt. Recent advances in our ability to decipher the warning signs have led to a number of successful evacuations. Three months before the dramatic eruption of Mount Pinatubo in the Philippines in June 1991, for example, scientists detected tremors on its flanks. Soon after, the volcano started steaming and puffing out clouds of ash. As activity increased, the government ordered an evacuation of 60,000 people, saving thousands of lives. In November 2017 tremors and ash clouds from Mount Agung on the Indonesian island of Bali triggered a similar evacuation of 40,000 people.

Although not all volcanoes give clear signals, even the smallest of signs can now be used to predict eruptions. Measuring the swelling of volcanoes has got easier with sensitive tilt meters and GPS sensors. Subtle changes in the sound of the ocean were successfully used to forecast the eruption of Piton de la Fournaise on the island of Réunion in the Indian Ocean in July 2006 and April 2007. Scientists monitoring the low-frequency seismic waves generated by the ocean hitting the sea floor had noticed that, when an eruption was imminent, sound waves passing through magma chambers slowed down. Based on this observation, local people were evacuated with several days' warning.

Even keeping an eye on the weather could also aid predictions. Pavlof, an active volcano on the Alaskan peninsula, is most active during the autumn and winter. One explanation is that storms at this time cause water levels to rise around the volcano, squeezing the magma up like toothpaste from a tube.

When the ground shakes

Our awareness of earthquakes dates back to our earliest days as a sentient species, but for most of human history we have not understood their causes. Only in the past century have scientists been able to answer the question: what exactly is an earthquake?

Quakes in the ancient world, including around the Mediterranean and Middle East, occurred frequently enough to have been part of the cultural fabric of early civilizations. Legends ascribing geophysical unrest to the whims and fancies of spiritual beings are a recurring theme in early cultures. In more recent history, people began to look for physical explanations. The ancient Greeks and Romans in the shape of Aristotle and Pliny the Elder, for example, proposed that earthquakes happened as a result of underground winds.

The earliest scientific studies of earthquakes date back to the eighteenth century, sparked by an unusual series of five strong earthquakes in England in 1750 followed by the Great Lisbon earthquake of 1755 in Portugal. Early investigations included cataloguing past earthquakes and trying to understand the seismic waves of energy that were generated during the events. These waves, which radiate from the earthquake's source and cause the ground to heave, remained the focus of scientific efforts until the end of the nineteenth century.

Following the 1891 Mino–Owari earthquake – the strongest inland quake ever to hit Japan – and the 1906 San Francisco earthquake, attention shifted to the mechanisms that give rise to these events. Using data from triangulation surveys – an early forerunner of GPS – conducted before and after the 1906 earthquake, geophysicist Harry Fielding Reid developed one of the basic tenets of earthquake science, the theory of 'elastic rebound'. This describes how earthquakes occur as a result of the abrupt release of stored stress along a fault line (see Figure 4.4).

Another half-century elapsed before the plate tectonics revolution of the mid-twentieth century provided an explanation for the more fundamental question: what drives earthquakes?

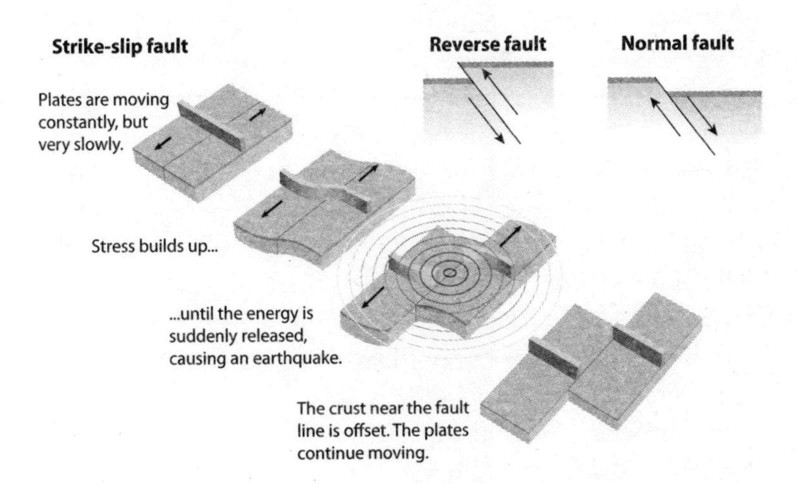

FIGURE 4.4 The elastic rebound theory describes how earthquakes happen at faults because of the movement of plates.

We now know that most earthquakes are caused by the build-up of stress along the planet's active plate boundaries, where tectonic plates converge or slide past each other.

Other earthquake causes have also been identified, such as post-glacial rebound, when the crust returns to its non-depressed state over timescales of tens of thousands of years following the retreat of large ice sheets. Such processes, however, make up only a tiny percentage of the overall energy released by earthquakes due to plate tectonics.

Sizing up quakes

By the early twentieth century geologists knew that some earthquakes create visible rips across Earth's surface, which gives some indication of their force. But since most fault ruptures are entirely underground, we need other methods to size up and compare earthquakes.

The earliest scales were called intensity scales, which typically assign Roman numerals to the severity of shaking at a given location. Intensity scales remain in use today: well-calibrated intensity values determined from accounts of earthquake effects help us study historical earthquakes and their effects within densely populated areas, for example. Following an earthquake in Virginia in 2011, more than 140,000 people reported their accounts to the United States Geological Survey's 'Did You Feel It?' website.

To size up an earthquake directly, one needs to record and dissect the waves it generates. Today this is done with seismometers employing digital recording, but it wasn't always so. The first compact instrument capable of faithfully recording small local earthquakes was called a Wood–Anderson seismometer. When the ground shook, a mass suspended on a tense wire would rotate, directing a light on to photosensitive film. The image 'drawn' by the light reflected the severity of the seismic waves passing through.

In the early 1930s Charles Francis Richter used these seismometers to develop the first magnitude scale – borrowing the word 'magnitude' from astronomy. Richter's scale is logarithmic, with each unit increase in magnitude corresponding to a 30-fold increase in energy release. So, a magnitude-7 earthquake releases almost 1000 times the energy of a magnitude-5 earthquake.

Magnitude values are relative: no physical units are attached. Richter tuned the scale so that magnitude 0 (M0) was the smallest earthquake that he estimated could be recorded by a surface seismometer under ordinary conditions. Earthquakes with negative magnitudes are possible but are unlikely to be recorded. The scale is open-ended, but Richter might have had an upper limit of M10 in mind: he also tuned the scale so that the largest recorded earthquakes in California and Nevada were around M7, and surmised that the 1906 San Francisco earthquake was probably around M8. (The largest earthquake recorded since then was the Valdivia earthquake in Chile in 1960, with an estimated magnitude of 9.5.)

Relationships have been developed since to relate the energy released by earthquakes to magnitude. In the 1960s Keiiti Aki introduced a fundamentally different quantity: the 'seismic moment'. This provides a full characterization of the overall size of an earthquake and is the measure generally used in scientific analyses.

What's called the moment-magnitude scale was introduced to convert the seismic moment to an equivalent Richter magnitude. This figure is the one usually reported in the media. Strictly speaking, this reported value is not 'on the Richter scale', because it is calculated differently from Richter's formulation. Still following Richter's approach, moment-magnitude values have no physical units, and are useful for comparing earthquakes.

Shake, rattle and roll

For assessing seismic hazards, the study of ground motion is 'where the rubber meets the road'. If we understand the shaking, we can design structures and infrastructures to withstand it. The severity of earthquake shaking is fundamentally controlled by three factors: earthquake magnitude, the attenuation of energy as waves move through the crust, and the modification of shaking due to the local geological structure.

FIGURE 4.5 Ground motion during an earthquake is what tends to do most damage. If we can understand it, we might be able to protect against it.

Bigger earthquakes generally create stronger shaking, but not all earthquakes of a given magnitude are created equal. Shaking can depend significantly on factors such as the depth of the earthquake, the orientation of a fault, whether or not the fault break reaches the surface and whether the earthquake rupture is relatively faster or slower than average.

Attenuation of seismic waves varies considerably in different regions. In a place like California or Turkey, where the crust is relatively hot and highly fractured, waves dissipate – or attenuate – quickly. Following the 1906 San Francisco earthquake, pioneering geologist G. K. Gilbert observed: 'At a distance of twenty miles [from the fault] only an occasional chimney was overturned ... and not all sleepers were wakened.' In regions that are far from active plate boundaries, such as peninsular India or the central and eastern USA, waves travel far more efficiently. The three principal mainshocks of the 1811–12 earthquake sequence around New Madrid in present-day Missouri damaged chimneys and woke most sleepers in Louisville, Kentucky, some 400 kilometres away (see Chapter 5). In 2011 the magnitude-5.8 Virginia earthquake was felt in Wisconsin and Minnesota, more than 1500 kilometres away.

Local geological structures such as soft sediment layers can amplify wave amplitudes. For example, the M8 earthquake along the west coast of Mexico in 1985 generated a ringing resonance in the lake-bed sediments beneath Mexico City. And in Port-au-Prince, some of the most dramatic damage in the 2010 Haiti earthquake was associated with amplification by small-scale topographic features such as hills and ridges.

Characterization of the full range and nature of the response of different sites remains a prime target for ground motion studies, in part because of the potential to map out the variability of hazard throughout an urban region, called microzonation.

This offers the opportunity to identify urban areas that are more and less hazardous, which can guide land-use planning and appropriate building codes.

Tsunami!

Undersea earthquakes can generate a potentially lethal cascade: a fault break can cause movement of the sea floor, which displaces the water above to set off a tsunami wave.

Tsunamis can also be generated when earthquakes trigger undersea slumping of sediments, although these waves are generally more modest in size.

Tsunami waves spread out through the ocean in all directions, travelling in the open ocean about as fast as a jet plane. They have a very long wavelength and low amplitude at sea, but grow to enormous heights as the wave energy piles up against the shore.

Connected quakes

Earthquakes are often related to one another – one can lead to another – but there are common misconceptions about what drives them and the ways they are linked.

It is an enduring misperception that a large earthquake is associated with a sudden lurching of an entire tectonic plate. If one corner of the Pacific plate moves, shouldn't it be the case that other parts of the plate will follow suit? The idea might be intuitive, but it is wrong. Earth's tectonic plates are always moving, typically about as fast as human fingernails grow. What actually happens is that adjacent plates lock up, causing warping of the crust and the storing up of energy, but only over a narrow zone along the boundary. So when an earthquake happens, this kink is catching up with the rest of the plate.

Earthquake statistics do tell us, however, that the risk of aftershocks can be substantial: on average, the largest aftershock will be about one magnitude unit smaller than the mainshock. Aftershocks cluster around the fault break, but can also occur on close neighbouring faults. As the citizens of Christchurch, New Zealand, learned in 2011, an M6.1 aftershock had far worse consequences than the M7 mainshock, because the aftershock occurred closer to a population centre.

In addition to aftershock hazard, there is always a chance that a big earthquake can beget another big earthquake nearby, typically within tens of kilometres, on a timescale of minutes to decades. For example, the 23 April 1992 M6.1 Joshua Tree earthquake in southern California was followed by the 28 June 1992 M7.3 Landers earthquake, approximately 35 kilometres to the north. Such triggering is understood as a consequence of the stress changes caused by movements of the rocks. Basically, motion on one fault will mechanically nudge adjacent faults, which can push them over the edge, so to speak.

Another mechanism is now recognized as giving rise to triggering: the stress changes associated with seismic waves. Remote triggering occurs commonly – but not exclusively – in active volcanic and geothermal areas, where underground magmatic fluid systems can be disrupted by passing seismic waves.

Overwhelmingly, remotely triggered earthquakes are expected to be small. Here again, advances in earthquake science as well as centuries of experience tell us that earthquakes do not happen in great apocalyptic cascades. However, in recent decades scientists have learned that faults and earthquakes communicate with one another in far more diverse and interesting ways than the classic foreshock–mainshock–aftershock taxonomy suggests.

Have we cracked earthquake prediction?

When seismologists are asked whether earthquakes can be predicted, they tend to quickly answer no. Sometimes even geologists can forget that, in the ways that matter, earthquakes are too predictable. We know where in the world they are likely to happen. For most of these zones, we have quite good estimates of the expected long-term rates of quakes. And while we often cannot say that the next Big One will strike in a human lifetime, we can say that it is very likely to occur within the lifetime of a building.

We know the largest earthquakes happen along subduction zones, with rupture lengths of more than 1000 kilometres and an average slip along a fault of tens of metres. But any active plate boundary is a potential site for a big earthquake, at any time. For example, two years before the 2010 earthquake in Haiti, geophysicist Eric Calais and his colleagues published results of GPS data from the region, noting that 'the Enriquillo fault is capable of a M7.2 earthquake if the entire elastic strain accumulated since the last major earthquake was released in a single event'. While this exact scenario did not play out in 2010, it wasn't far off. We can say for sure that people living on plate boundaries will always face risk.

Future large earthquakes are expected in California. Research by James Lienkaemper and his colleagues estimate that sufficient strain is stored on the Hayward Fault in the east San Francisco Bay area to produce a M7 earthquake. An earthquake this size is expected, on average, every 150 years. The last one was in 1868. Knowing such information inevitably increases local anxieties, but earthquakes occur by irregular clockwork: if the average repeat time is 150 years, it could vary between 80 to 220 years. So, we are left with the same vexing uncertainty: an 'overdue' earthquake might not occur for another 50 years, or it could happen tomorrow. On a geological timescale there is not much difference between sooner versus later. On a human timescale, however, sooner versus later seems like all the difference in the world.

Earth scientists have made great strides in forecasting the expected average rates of damaging earthquakes. The far more challenging problem remains finding the political will and resources to prepare for the inevitable.

Why prediction is so difficult

In the 1970s and 1980s leading scientists were quoted in the media expressing optimism that reliable short-term prediction of earthquakes was around the corner. This was fuelled by promising results from the Soviet Union, and the apparently successful prediction of the 1975 earthquake in Haicheng, China. Since then, this optimism has given way to varying degrees of pessimism. Why are earthquakes so hard to predict?

Any number of possible precursors to earthquakes have been explored: small earthquake patterns, electromagnetic signals, radon releases and hydrogeochemical changes. Many of these seemed promising, but none have stood up to rigorous examination.

Consider this example. In March 2009 Italian laboratory technician Giampaolo Giuliani made a public prediction that a large earthquake would occur in the Abruzzo region of central Italy. His evidence? An observed radon anomaly. The prediction was denounced by local seismologists. On 6 April the M6.3 L'Aquila earthquake struck the area, killing 308 people.

This gets to the issue of reliable precursors. It is possible that radon was released because of the series of small earthquakes, or foreshocks, that preceded the main earthquake. It is also possible that it was coincidence. Scientists explored radon as a precursor in the 1970s and quickly discovered how unreliable it is. Once in a while radon fluctuations might be associated with an impending earthquake, but usually they are not. Meanwhile, big earthquakes hit regions where radon anomalies did not take place. The same story has played out with many other proposed precursors.

That's not to say that seismologists have neglected to investigate precursors – on the contrary, they are examining them with increasingly sophisticated methods and data. However, a common bugaboo of prediction research is the difficulty of truly prospective testing. To develop a prediction method based on a particular precursor, researchers compare past earthquakes with available recorded data. One might, for example, identify an apparent pattern of small earthquakes that preceded the last ten large earthquakes in a given region. Such retrospective analyses are plagued by subtle data selection biases. That is, given the known time of a big earthquake, one can often look back and pick out apparently significant signals or patterns.

This effect is illustrated by the enduring myth that animals can sense impending earthquakes. It is possible that animals respond to weak initial shaking that humans miss, but any pet owner knows that animals behave unusually all the time – and it's soon forgotten. People only ascribe significance with hindsight.

At present, most seismologists are pessimistic that prediction will ever be possible. But the jury is still out. One of the big unanswered questions in seismology is: what happens inside

Earth to set an earthquake in motion? It is possible that some sort of slow nucleation process is involved, and therefore possible that earthquake precursors exist. For this, as well as all earthquake prediction research, the challenge is to move beyond the retrospective and the anecdotal, into the realm of statistically rigorous science.

5
Shifting ideas

Today plate tectonics is the orthodoxy, but it's not ideal. The theory struggles to account for earthquakes far from plate boundaries, and does not extend to events deep inside Earth that affect the surface. Our present knowledge also makes it difficult to predict how our continents will move in future. Fortunately, there is plenty of new thinking aimed at fixing these imperfections.

Quakes in the 'wrong' places

Just over 200 years ago, between 16 December 1811 and 7 February 1812, a series of four massive quakes ripped through the Mississippi embayment, a low-lying, sediment-filled basin stretching from the Gulf of Mexico northwards to Cairo, Illinois. Centred on the town of New Madrid in present-day Missouri, the quakes measured around magnitude 7, and possibly hit 8. In the last of them, the Mississippi River flowed backwards, the riverbanks spewed sand, and Reelfoot Lake – today a popular hunting and fishing preserve in north-west Tennessee – formed as the ground opened to swallow displaced water (see Figure 5.1).

FIGURE 5.1 Reelfoot Lake in Tennessee formed after an earthquake that made the Mississippi flow backwards.

That, on the face of it, is rather unexpected. New Madrid lies far from typical arenas of major seismic upheaval – the boundaries between tectonic plates. But such earthquakes in the middle of plates were not unique. In 1556 the deadliest earthquake on record occurred in Shaanxi Province in China's northern interior, again nowhere near a plate boundary. Some 800,000 people were killed as, according to a contemporary report, 'mountains and rivers changed places'. On 23 August 2011 a magnitude-5.8 quake struck with an epicentre near Mineral, Virginia. There were no deaths, but the incident caused chaos and confusion up and down the US east coast. Earthquakes have struck the interiors of India and Australia in the recent past as well.

These 'intraplate' earthquakes have long been a mystery. And what we are finding out is giving us pause for thought. It might be that it's not just San Francisco and Los Angeles that are susceptible to significant quakes, but maybe New York, Sydney and London, too.

San Francisco and Los Angeles are typical of earthquake-prone cities. They sit near the infamous San Andreas Fault in California, where the North American and Pacific plates grind against each other at a rate of 33 to 37 millimetres a year, building up strains that are released in earthquakes.

It now seems that the forces that rip Earth's crust apart at plate boundaries and crash them back together may also be at work in intraplate earthquakes; it's just that full-thickness rips never quite happen. The result is an unstable region that, though often unremarkable at the surface, is more easily stressed than the rock around it. Strains build up more slowly, which could explain why quakes here happen far less frequently than those at plate boundaries.

In the 1980s it became clear that New Madrid sits atop such a 'failed rift'. Dubbed the Reelfoot Rift, it lies buried beneath the Southern and Midwestern USA and seems to have shuddered regularly in recent millennia. Evidence to support this comes from analysis of geological features called sand blows, produced when a powerful earthquake shakes the soil so much that it loses strength and behaves like a liquid, spewing from the ground in a tiny mud volcano. The plains around New Madrid are dotted with sand blows that formed 200 years ago. Underground, there are more, suggesting that large tremors racked the area in the years 300, 900 and 1450.

The United States Geological Survey suggests that there is a 25 to 40-per-cent chance of a magnitude-6 or larger quake hitting the New Madrid area in the next 50 years, with a 7 to 10-per-cent chance of an event as big as the one two centuries ago. Back then, there were hardly any settlers in the region. Today a quake that size would play havoc with the area's much larger population. New Madrid might not be the only area at risk. Studies of the deformation of sediments beneath the Mississippi River have uncovered a 45-kilometre-long fault north of Memphis that seems to be part of the Reelfoot system. Then there's the 10-kilometre-long Marianna Fault in Arkansas, discovered in 2009. The seismogenic potential appears to involve a much larger area than just the active faults we see today.

Yet now may not be a time to worry. If the faults in the area are still under strain, they should be moving, just as they are at the San Andreas Fault. Yet 20 years of GPS studies of the seismic zone around New Madrid shows that they are not. In 2009 the author of those studies, Seth Stein of Northwestern University in Evanston, Illinois, and his colleague Eric Calais suggested that New Madrid is now in a deep seismic slumber from which it should not be expected to awake for hundreds, if not thousands, of years.

Migrating quakes

This leads Stein to make a controversial claim. He doesn't buy the idea that intraplate earthquakes are akin to interplate quakes, hitting home less frequently but in similarly predictable places. Instead, he characterizes them as episodic, clustered and migrating: seismic energy can jump within a network of small faults that snake their way through the middle of a tectonic plate. Beneath the US Midwest he reckons that over time the motion in New Madrid will be transferred into seismic zones in Indiana and further south into Arkansas.

Work by Mian Liu of the University of Missouri in Columbia supports this picture. Liu analysed the occurrence of intraplate earthquakes over 2000 years in the north of China and showed that the epicentres of intraplate earthquakes in China hop around haphazardly. Areas of violent shocks become quiescent; previously docile areas suddenly become active. For him, earthquakes appear to be 'spatially migrating, jumping from one fault to another across long distances'. Faults in the middle of a plate, it seems, may be mechanically coupled, so that an earthquake along one changes another's susceptibility to future movement.

If so, that could have huge ramifications for our understanding of intraplate quakes. Take the Virginia quake of 2011. Its epicentre was in the Central Virginia seismic zone, which has experienced many quakes of around magnitude 3 over the past 120 years, but was not considered particularly at risk of anything bigger. If Stein and Liu's ideas are right, the culprit might have been seismic energy that roamed into the area from elsewhere. The nearby Western Quebec seismic zone, for example, extends over the northern border of New York State, and was visited by a magnitude-5.6 earthquake in 1944. The Eastern Tennessee seismic zone, stretching from north-east Alabama to south-west Virginia, is also highly active. Two

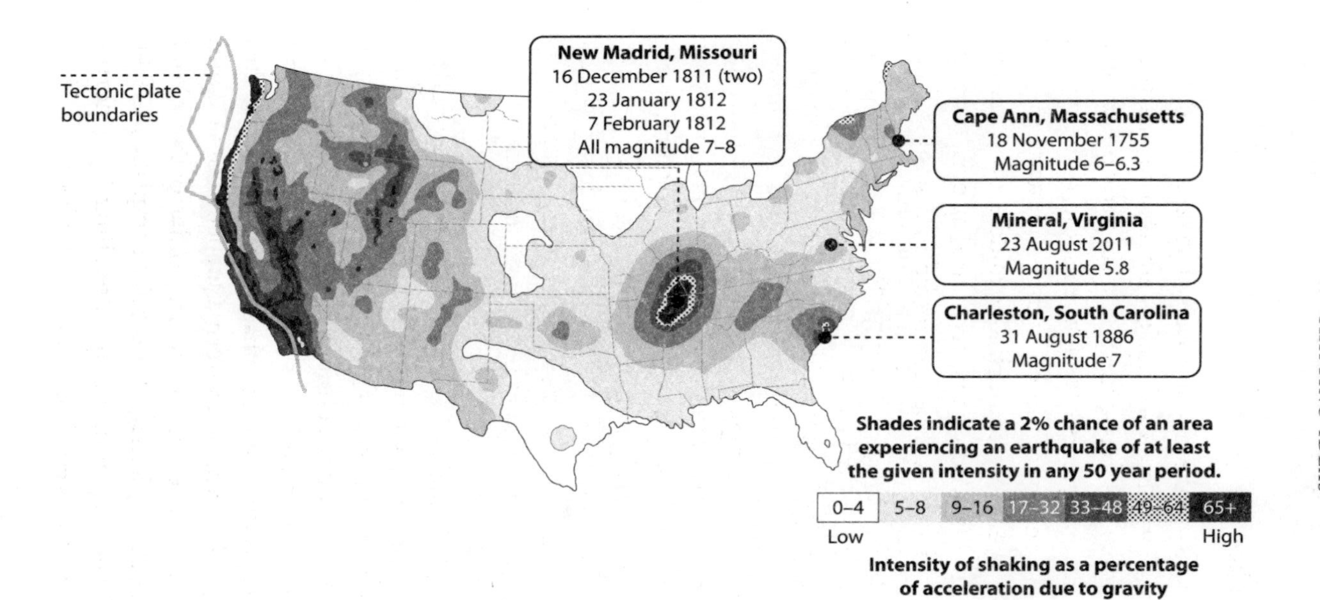

FIGURE 5.2 Seismic risk maps of the US mainland based on historical data show that places well away from plate boundaries may still be prone to earthquakes.

magnitude-4.6 earthquakes have occurred there in recent decades.

That amounts to a wake-up call. Earthquakes like the ones in Virginia and New Madrid could happen anywhere, including in Boston, Chicago, New York and other major cities. An earthquake of estimated magnitude 5.7 hit the Dover Strait off south-east England in 1580, causing a pinnacle to fall off Westminster Abbey in London some 150 kilometres away. A magnitude-4.3 quake struck the same region in 2007. We should not overstate the risks: most modern buildings in Western cities could easily withstand a magnitude-5 or 6 quake. Skyscrapers in particular have enough 'sway' in them to counteract the effects, but historical monuments and older buildings made from unreinforced brick could be vulnerable. Critical infrastructure such as electrical and telecommunications networks, water and fuel pipelines may also be at risk.

What seismologists would dearly love to find is some kind of order to the behaviour of intraplate earthquakes. In the USA, almost every third state is thought to have a piece of failed rift in it. Why some like the Reelfoot are seismically active but others are not remains a big unanswered question. Without a clear pattern to explain intraplate quakes, it seems we have to expect that they could happen anywhere, at any time.

The power of plumes

It's not just earthquakes that appear 'in the wrong places'. Volcanoes do the same. US geophysicist Jason Morgan was a pioneer of plate tectonics (see Chapter 4), but in the 1970s he was also one of the first to find fault with the theory's explanation for one particular surface feature, the volcanism of the Hawaiian islands.

These islands lie thousands of kilometres from the edge of the Pacific plate on which they sit. According to plate tectonics, all the significant action happens at those boundaries. So the plate-tectonic explanation for these islands is that their volcanism is caused by a weakness in the plate that allows hotter material to well up passively from the mantle. But, reviving an earlier idea of the Canadian geophysicist John Tuzo Wilson, Morgan suggested instead that a plume of hot mantle material is actively pushing its way up from many thousands of kilometres below and breaking through to the surface.

That went against the accepted theory, and it wasn't until the mid-1980s that others began to think Morgan might have a point. The turnaround came when seismic waves unleashed by earthquakes began to reveal some of our underworld's structure.

FIGURE 5.3 The Hawaiian islands are volcanic, yet they are nowhere near a plate boundary. How do we explain that?

The seismic images are rough and fuzzy, but seem to reveal a complex, dynamic mantle. Most dramatically, successive measurements have exposed two massive regions of very hot, dense material – dubbed thermochemical piles – sitting at the bottom of the mantle near its boundary with the outer core. One is under the southern Pacific Ocean, and one beneath Africa. Each is thousands of kilometres across, and above each a superplume of hotter material seems to be rising towards the surface.

A superplume could explain why the ocean floor in the middle of the southern Pacific is raised some 1000 metres above the surrounding undersea topography, another thing plate tectonics has difficulty explaining. Something similar goes for the African superplume, which is propping up an area from south of the Congo all the way down to southern South Africa, including Madagascar. Seismic imaging reveals smaller plume-like features extending upwards towards the lithosphere beneath Iceland and Hawaii – perhaps explaining both these islands' existence and their volcanism.

Off the coast of Argentina, meanwhile, the sea floor plunges down almost a kilometre, directly above a mantle region that seismic imaging identifies to be cold and downwelling. Likewise, the Congo basin lies on a cold area and is hundreds of metres lower than its surroundings. Almost everywhere we look, there is evidence of vertical movements within Earth reshaping its surface.

Deep action

What is less clear is what mechanisms are at work. Standard plate tectonics has it that material plunging downwards at subduction zones is recycled in the shallow mantle, reappearing through volcanic activity nearby or further afield at boundaries where two plates are being pushed apart. Blurry yet tantalizing seismographs, however, show sections of subducted plates at various stages of descent through Earth's interior towards the lower mantle (see Figure 5.4).

Simulations by Bernhard Steinburger, now at the University of Oslo, Norway, and his colleagues show how a subducted slab, once it arrives at the boundary between the mantle and the core, can bulldoze material along that layer. When this material meets a thermochemical pile, plumes begin to form above. Steinberger's modelling shows plumes developing at more or less the right places. For example, their model shows that slabs being subducted beneath the Aleutian Islands near Alaska could

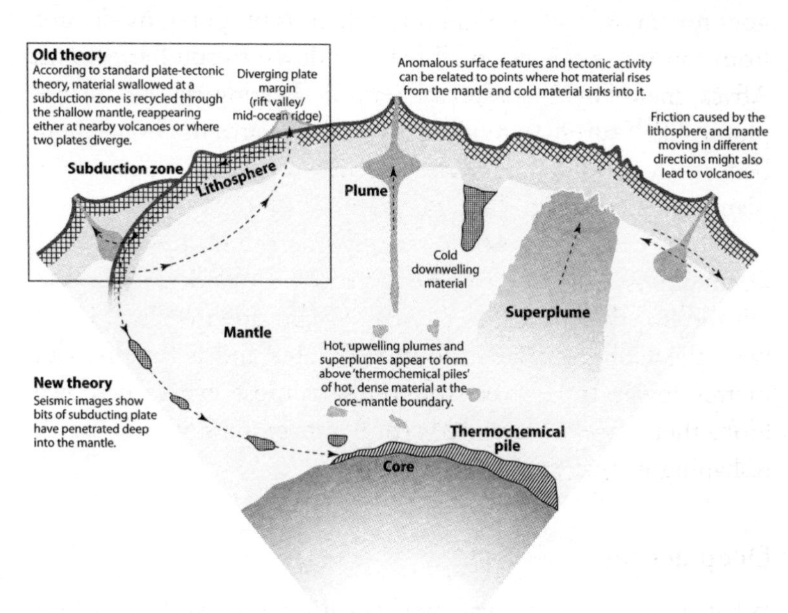

FIGURE 5.4 Seismographic images suggest that the workings of the deep Earth have an important effect on surface features.

trigger a plume beneath Hawaii, creating a hotspot that fuels the Hawaiian volcanoes.

Clint Conrad at the University of Hawaii at Manoa and his colleagues have modelled the effect of a tectonic plate moving one way while the mantle beneath moves in the opposite direction. They found that in certain cases the resulting shearing effect can cause mantle material to melt and rise.

This model accurately predicts that volcanic seamounts should be present on the west but not the east of the East Pacific Rise, a mid-ocean ridge that runs roughly parallel to the western coast of South America. Seismic measurements indicate that the mantle and the plate to the west are moving in opposite directions; to the east they are not. The model also predicts that the shearing effect is largest under the western USA, southern Europe, eastern Australia and Antarctica – all areas of volcanic activity away from plate boundaries.

More than plate tectonics

If the dynamics of the deep Earth can change surface topography today, the same must have been true in the past. But as we have seen, deciphering the history of Earth in the distant past is difficult indeed.

In 2011 geologist Nicky White at the University of Cambridge, UK, and his colleagues found some clues to a small part of the story off the west coast of Scotland. They set off explosions and recorded the reflections to get a sense of what lies beneath the sea floor. What they saw buried under more recent layers of rock and sediment were fossil landscapes some 55 million years old, replete with hills, valleys and networks of rivers.

By analysing how these rivers had changed course over time, the team showed that the region was once pushed almost a kilometre above sea level before being buried again, all in the space of a million years. That is far too quick for plate tectonics to throw up a mountain range and have erosion wear it down again. Instead, White points to a blob of hot mantle material that he says travelled radially outwards from the mantle plume that is possibly fuelling the volcanoes in nearby Iceland.

Other researchers have identified similarly precipitous vertical movements of land that is now in eastern Australia during the Cretaceous period. Again, the short timescales pretty much discount simple plate tectonics. The convecting mantle looks a more likely instigator.

Even iconic events of Earth's tectonic past might not be all they seem. The Himalayas had formed by 35 million years ago, after the Indian plate sped north and slammed into the Eurasian plate. But plate tectonics struggles to explain why India zoomed towards its target at speeds of up to 18 centimetres a year. Today plates reach speeds of only about 8 centimetres per year. One controversial explanation is that its headlong motion was fuelled by the mushrooming head of a mantle plume.

Likewise, the anomalous and periodically devastating seismicity of the US Midwest – far from any plate boundaries – might be explained by the motion of the ancient Farallon plate, which started slipping into the mantle along the west coast of North America during the Cretaceous. By the early nineteenth century, that plate had burrowed deep enough to cause a downwelling below the mid-Mississippi river valley, deforming the overlying lithosphere sufficiently to trigger the series of earthquakes centred on New Madrid.

Not everyone is convinced by the role of plumes. The idea's biggest hole is that while seismic waves do travel more slowly beneath supposed hotspots in the shallow mantle, these velocity anomalies don't extend all the way down to the bottom of the mantle where the plumes are thought to form. Enthusiasts for a deeper explanation of Earth's surface activity think it is only a matter of time and better seismic imaging before these objections are also countered.

If those enthusiasts are correct, more than plate tectonics is at work shaping our planet's past, present and future. In that case, geology is on the cusp of another revolution which could be as significant as plate tectonics.

The next supercontinent

Plate tectonics drives the slow-grinding motions of our continents. Armed with our understanding of its actions, you might think we could predict the future movements of our great landmasses. Indeed, there is a band of fortune tellers who do exactly that. The problem is that the members of this group all have different ideas about where the continents will end up, and why. Our inability to agree on the shape of things to come implies that we are still missing vital details about plate tectonics.

Magnetic signals recorded in sea-floor rocks, and chemical traces from the roots of ancient mountain ranges, tell us how continental drift has changed the face of Earth. They point clearly to a time 180 million years ago when all today's continents were stuck together in one vast landmass centred roughly where present-day Africa is: the supercontinent Pangaea (see Chapter 2).

We know that Pangaea came together about 330 million years ago. Before that, some consider a relatively short-lived gathering of continents near the South Pole to be another supercontinent, named Pannotia or Greater Gondwana. Going back further, another supercontinent, Rodinia, probably dominated the planet between about 1.2 billion and 700 million years ago. And about 2 billion years back it is thought that there was yet another.

The implication is that in future we can expect another supercontinent. But just how and where will this vast landmass form?

Novopangea

One model simply projects what's happening today into the future. The great split that broke Pangaea apart is still growing, with Africa and Eurasia moving away from the Americas. The Atlantic is spreading as new rock wells up at its mid-ocean ridge, while the Pacific is shrinking, consumed by the subduction zones that surround it, the famous ring of fire.

If these motions continue, in about 250 million years a new supercontinent, Novopangaea, would form on the opposite side of Earth from the original Pangaea, as the Americas and Asia crunch in around northbound Australia (see Figure 5.5).

But it may not be that simple. Christopher Scotese at Northwestern University in Evanston, Illinois reckons projecting 50 million years into the future might work, but to see further ahead takes a detailed understanding of plate tectonics that we are still struggling to achieve.

Since 1982 Scotese has been making maps of past and future Earth using various rules of thumb. The most important rule is widely accepted: plate tectonics is driven mainly by the pull of sinking slabs at subduction zones, with a smaller push from new rock forming at mid-ocean ridges. Work out the layout of subduction zones and ridges at any time, and you can begin to see how the continents should be tugged and nudged around.

But three kinds of cataclysmic event can change the course of this smooth voyage. A subduction zone can swallow a spreading ocean ridge, as is happening today off the west coast of North America, where the Juan de Fuca ridge is being consumed.

Or two unsinkable continents can collide, snuffing out a sandwiched subduction zone and forcing up great mountains. This is exactly what India and Eurasia did to build the Himalayas.

The third possible cataclysm is much harder to fathom or predict: the creation of a new subduction zone. This must happen somehow or all existing subduction zones would eventually be killed off by colliding continents, and plate tectonics would cease altogether. But evidence that plates have been moving for much of Earth's history, with many cycles of supercontinent formation and destruction, indicates that new subduction zones must start up.

Pangea Proxima

The most likely spot for new subduction is at passive margins, for example on the Atlantic coasts of Europe, Africa and the Americas. These are places where old oceanic lithosphere, spreading out from the mid-ocean ridge, meets continental crust. The oceanic lithosphere has had time to cool since it formed and become denser than the rock beneath, so it wants to sink.

But it can't. Old, cold lithosphere rocks are hard to crack. Even the weight of kilometres-deep river sediments washed on to the passive margin from the continents isn't enough on its own. The weakening effect of water seeping into the rocks may help, but probably not enough to crack those passive margins. In the 1980s Scotese suggested that the secret must be a 'nick' in the oceanic lithosphere, where localized stresses help it to tear more easily.

There are already two small subduction zones in the western Atlantic: the Lesser Antilles volcanic arc, which forms the eastern boundary of the Caribbean Sea, and the Scotia arc between the tip of South America and the Antarctic Peninsula. Scotese

Africa · North America · Australia · South America · Antarctica · Eurasia · New land

Pangea proxima

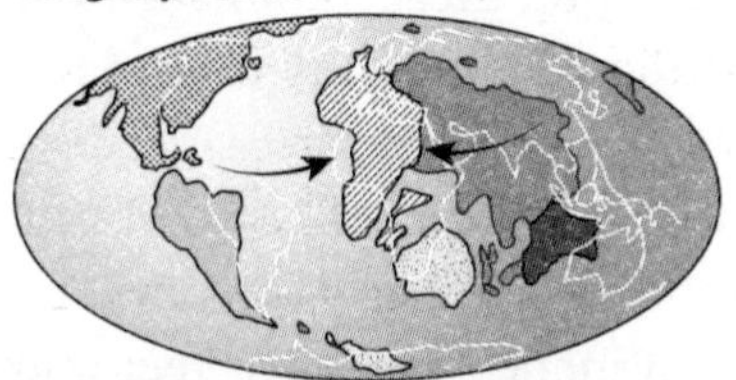

150 million years

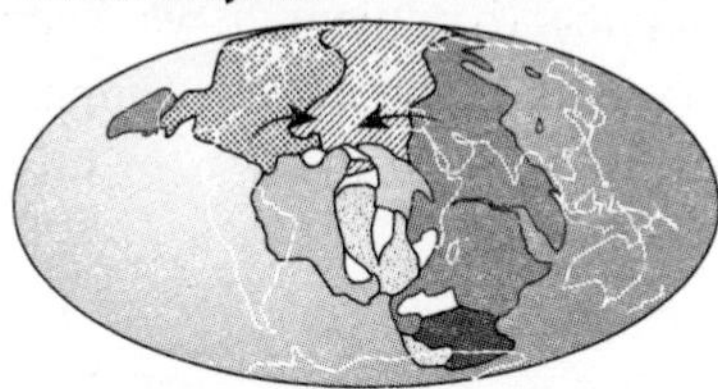

300 million years

In around 100 million years, subduction spreading along the western side of the Atlantic causes it to start closing. North America ends up fused with the west cost of Africa, with South America swinging round to end up at the south of a new supercontinent centred on the present-day Atlantic.

Aurica

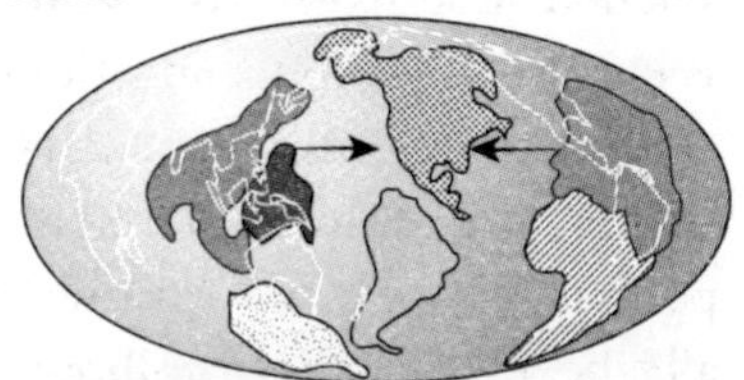

150 million years

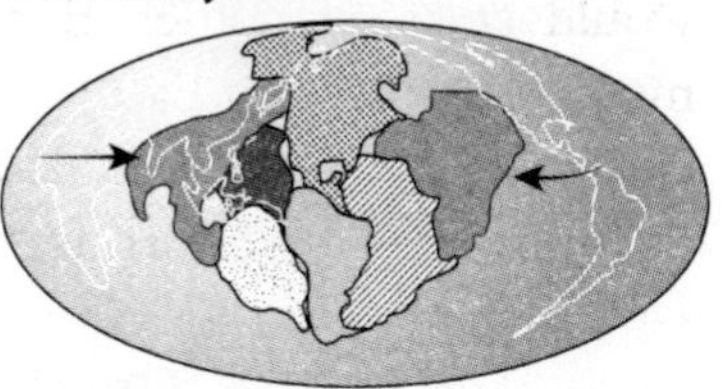

300 million years

Subduction starts on both sides of Atlantic, and both the Atlantic and the Pacific close. Eurasia splits, its western half moving westwards with Africa, and its eastern half migrating eastwards. It scoops up Australia to form a new supercontinent, centred where the Pacific is now.

Amasia

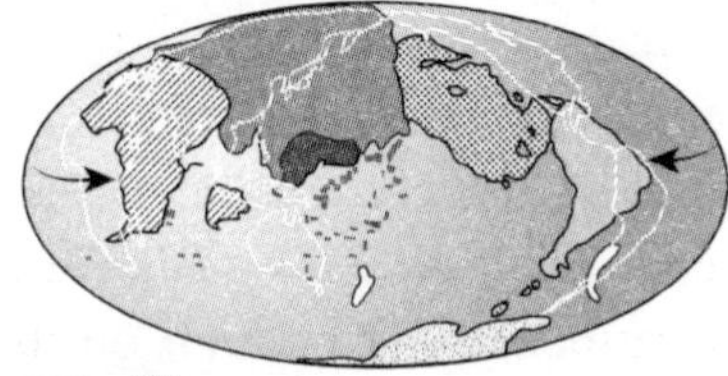

250 million years

The Atlantic continues to open but widens more at the south. Africa moves west and Australia moves north; meanwhile South America pivots and ends up with its west coast fused to North America's east cost. Antarctica remains aloof from this straggly new supercontinent.

Novopangaea

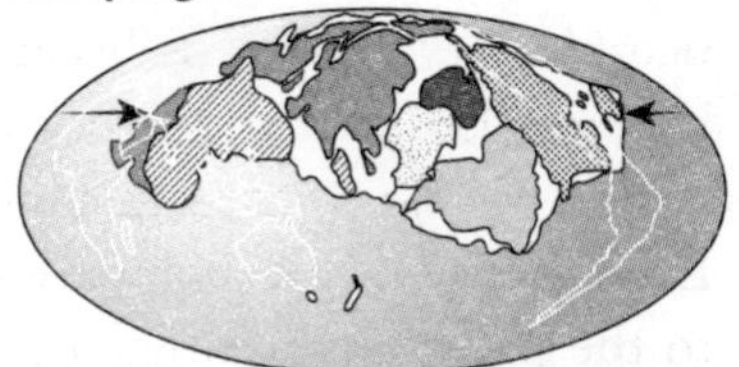

250 million years

Current tectonic movements continue, with the Atlantic widening and the Pacific consumed by subduction. South America swings westwards and northwards, scooping up Antartica and Australia; Africa rotates anticlockwise, taking Western Europe with it, its current south ending up fused with Arabia.

FIGURE 5.5 Scientists have forecast at least four different potential shapes for the next gathering of continents.

suggests that eventually these will spread south and north, joining up to form an extended subduction zone along the east coast of the Americas. In his projection, this will eat the Atlantic's mid-ocean ridge about 100 million years from now, and the Atlantic will start to close again. After 250 million years, the Americas will have collided with an already merged Africa and Eurasia – as will Australia and most of Antarctica – to form what Scotese calls Pangaea Proxima.

Amasia

In 2012 Ross Mitchell, then at Yale University, and his group mapped out a third route. The shifting of mass associated with the formation of a supercontinent affects Earth's rotation, changing its spin axis relative to the solid body of the planet. By looking at the orientation of magnetic crystals in rocks that cooled around the time that earlier supercontinents existed, the team showed that Rodinia formed about 90 degrees in latitude away from the position of the earliest supercontinent; and Pangaea coalesced about 90 degrees from Rodinia. Mitchell and his colleagues predict that the same thing will happen again, meaning that the next supercontinent should form somewhere near the North Pole, as Asia and North America crunch together. They call the result Amasia.

Aurica

Geoscientist João Duarte at the University of Lisbon, Portugal, thinks all these models have problems. Amasia and Novopangaea would both be surrounded by large areas of ocean crust more than 400 million years old, which he finds implausible. In 2008 Dwight Bradley at the United States Geological Survey in Anchorage, Alaska, looked at rocks around ancient passive margins and found the most aged were on average about 180 million years old. Hardly any lasted 400 million years. Duarte thinks this is no coincidence, but that plates in Atlantic-type oceans have to start subducting after about 200 million years.

Scotese's Pangaea Proxima does not have the old-crust problem: the Pacific could in theory stay open for many hundreds of millions of years with new crust constantly being generated and destroyed. But Duarte considers this improbable too, because ridges such as the Juan de Fuca are already being subducted.

Duarte agrees with Scotese that subduction may spread like a virus, a process he calls invasion. He has found evidence that subduction is beginning to invade the Atlantic's eastern margin, off the coast of Portugal, where forces generated by the remnants of ancient subduction in the Mediterranean are helping to create new faults in the ocean floor.

In Duarte's forecast, published in 2016, subduction will spread along both sides of the Atlantic within a few tens of millions of years, and the ocean will start to close. The Pacific will keep on closing, too, meaning something else has to give. That something is Asia. A rift cuts across the continent, from the Indian Ocean up to the Arctic, as the Himalayan Plateau collapses under its own weight. A new ocean opens up, and the eventual outcome is a new supercontinent with the two halves of Asia on the outside and American and Australia at its core – hence Aurica.

It's a nice try, says Scotese. But it has at least one difficulty. To close the Atlantic Ocean you have to subduct the mid-Atlantic ridge – but if you have subduction on both sides of the ocean, the ridge can stay in the middle and supply crust to both sides.

The proponents of all four ideas are keen to stress that the future is uncertain, and that their own model is just one option (although of course the most likely one). Whoever is right, our future descendants will inevitably have to adapt to a strangely shaped world, and will in turn be shaped by it.

6
Atmosphere, climate and weather

Gravity keeps our planet shrouded in a veil of gas that extends for hundreds of kilometres before merging with space. Towards its outer limit, the atmosphere displays curious magnetic and electrical properties, but most of our attention will be focused on the lowest 10 kilometres, where the air is breathable. This thin skin is home to life, it's where our weather happens, and it is a vital player in the chemical cycles that maintain Earth's geology and climate.

What makes an atmosphere?

Earth's atmosphere has changed dramatically over the planet's life. Any early gases would almost certainly have been swept away by outbursts from the newborn Sun. While it is possible that some gases arrived on Earth from cosmic visitors, the most likely source of the atmosphere is the gas spewed out by volcanoes. Today 64 per cent of volcanic emissions are water vapour, 10 per cent sulfur dioxide and 1.5 per cent nitrogen.

In the intervening millennia, the composition of this gaseous soup has changed beyond recognition. As the young Earth cooled, water vapour condensed to form the oceans, and vast quantities of carbon dioxide dissolved in the oceans and later formed vast beds of limestone. Throughout all this time, without anywhere else to go, inert nitrogen slowly built up.

What about oxygen? For 2 billion years, there was little or none in the atmosphere. That all changed when early life forms worked out how to perform photosynthesis using water. That single event, which precipitated the greatest mass murder Earth has ever seen, changed the atmosphere profoundly (see Chapter 8).

By 600 million years ago, the atmosphere had essentially the same make-up as now: 75 per cent nitrogen by mass, 23 per cent oxygen, 1 per cent water vapour and 1.3 per cent argon. There are traces – less than 0.1 per cent – of carbon dioxide, neon, helium and an array of other gases, from ozone and radon to hydrogen and nitric oxide.

From the bottom up

The lower atmosphere is formed of a series of layers of which the troposphere – the layer nearest to the ground – and the stratosphere, which sits above it, play key roles in determining patterns of weather and climate, and have the biggest direct influence on life.

The atmosphere gets its heat largely from the Sun in the form of electromagnetic radiation. (The amount of geothermal energy from hot radioactive rocks is so small it is usually ignored.) Most of the Sun's energy is radiated in the visible part of the spectrum, at wavelengths between 0.4 and 0.7 micrometres. This radiation passes through the atmosphere without being absorbed, and warms the surface of Earth.

About 7 per cent of the Sun's energy is radiated at wavelengths shorter than visible light, in the ultraviolet. This radiation is absorbed by molecules of oxygen and ozone in the stratosphere and warms that layer directly. A little solar energy

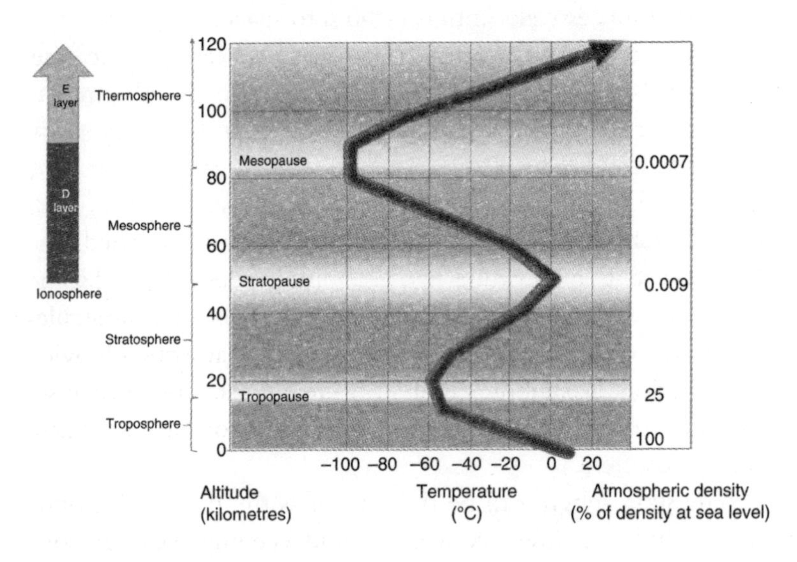

FIGURE 6.1 Earth's atmosphere forms a series of layers defined by rising or falling temperature.

is radiated at longer than visible wavelengths, in the infrared. Some of this is absorbed in the atmosphere, but plays only a minor part in keeping the air warm.

The Sun's energy reaches the atmosphere primarily from below, from the warm surface of Earth. This heating takes place partly through direct conduction of heat from the surface, but mainly because the planet radiates in the infrared part of the spectrum, and infrared radiation is absorbed by molecules such as water vapour and carbon dioxide in the lower atmosphere.

This infrared radiation makes the air warmer and, in turn, the air itself radiates heat, also at infrared wavelengths. Some of this radiation goes back down to the surface and keeps it warmer than it would otherwise be, creating what we now know as the greenhouse effect (see Chapter 10). The rest works its way upwards through the atmosphere, being absorbed and reradiated successively until it escapes to space.

The warmth of the troposphere causes heated air to rise, a key process in driving our weather and the overall circulation of the atmosphere. But this warm air can rise only so far because it is held down by warmer air in the stratosphere.

The stratosphere – which is pretty much synonymous with the ozone layer – is where ultraviolet radiation is absorbed. The radiation splits apart ordinary oxygen molecules (O_2) and some of the freed oxygen atoms react with other oxygen molecules to create ozone (O_3). The ozone itself also absorbs ultraviolet radiation, at slightly different wavelengths. Both processes extract energy from the passing solar radiation and so warm the stratosphere.

The stratosphere can be thought of as the lid of the troposphere, holding down convection and keeping weather confined. The average temperature at the planet's surface is about 15 °C, which is some 33 °C warmer than it would be if the

Earth had no blanket of air and no greenhouse effect to keep it warm. Rising through the troposphere, the temperature initially falls with increasing altitude, stopping at 20 kilometres. This is the boundary between the troposphere and the stratosphere: the tropopause. Its exact height varies with latitude, the seasons and from day to night. The troposphere contains about 75 per cent of the atmosphere of Earth by mass.

Saving ozone (and ourselves)

In May 1985 the world awoke to a disaster of potentially global proportions. Joe Farman of the British Antarctic Survey and colleagues announced that a huge hole had opened over the southernmost continent. The news caused a sensation because the hole was growing and stratospheric ozone is the planet's chief line of defence against harmful ultraviolet radiation, which causes skin cancer and cataracts. What's more, human beings were responsible. The discovery led to the world's most successful environmental treaty, so far.

The greediest ozone-eater was identified as chlorine, carried aloft in halocarbon compounds, such as chlorofluorocarbons (CFCs) and freons, used in industry as solvents and especially as the gases that kept our fridges cold. Once high in the atmosphere, chlorine escapes these compounds. On the surface of frozen particles in stratospheric clouds it breaks down ozone (O_3) into molecular oxygen (O_2).

In an astonishing display of unity, world governments came together and in 1987 agreed a ban on producing CFCs and other ozone-depleting chemicals under the aegis of the Montreal Protocol. It came into force in 1989 and has started to turn the tide. Satellite measurements showed ozone levels stabilizing in the mid-1990s and starting to recover in the 2000s.

Full recovery of ozone is not expected before 2065 and potential obstacles keep appearing that may delay this date. Hydrofluorocarbons, the main replacements for CFCs, turn out to be vigorous greenhouse gases, some 4000 times as potent as carbon dioxide. So, in 2016, new rules were agreed to phase these out. Then, in 2017, attention was drawn to rising levels of another ozone-eater not covered by the Montreal Protocol, called dichloromethane. Some estimates suggest that this compound could push back full recovery of stratospheric ozone to 2095.

Electric layers

From an altitude of about 20–60 kilometres, temperature increases through the stratosphere, from about −60 °C at the tropopause to a maximum of about 0 °C at the top of the stratosphere. About 99.5 per cent of the mass of the atmosphere lies below this point, the stratopause.

From 50 to 80 kilometres up another cooling layer, the mesosphere, occupies the now rapidly thinning atmosphere, with the coldest atmospheric temperatures, about −100 °C, reached at its highest point, the mesopause. From there on outwards, temperature increases in the last thermal layer, called the thermosphere. In this region ultraviolet and X-radiation are absorbed directly from the sun, and here atoms are ionized as electrons are knocked off them to leave positively charged ions behind.

Some of this ionization occurs at the top of the stratosphere, and everything from an altitude of 50 kilometres to 400 kilometres is regarded as the ionosphere. It has its own series of layers, each defined in terms of its differing degree of ionization. The presence of ionized particles in this region was first suspected when Guglielmo Marconi showed that radio waves can be transmitted 'around the corner' of the spherical Earth. Radio waves travel in straight lines, and the long-distance transmission of signals around the world – without the aid of satellites – is possible because ionized layers reflect radio signals at wavelengths longer than about 15 metres.

The lowest layer of the ionosphere, from 50 to 90 kilometres, is called the D layer and has only a low concentration of free electrons, and only reflects long-wavelength radio waves. Next up is the E layer, reaching to 150 kilometres. It is more strongly ionized than the D layer and reflects medium-wavelength radio waves. Its ionization can, however, disappear at night, which is why some radio signals come and go at different times. The F layer, from 150 to 400 kilometres, is the most strongly ionized region of the ionosphere and the most useful for radio communications.

Above an altitude of about 400 kilometres, the atmosphere is so tenuous that collisions between molecules, atoms and ions are too rare for it to be treated as a continuous gas, so the concept of temperature is no longer meaningful; in this region individual particles can escape into space, so it is sometimes referred to as the exosphere.

Magnetic shielding

Beyond the ionosphere, above about 500 kilometres, the magnetosphere is the outermost region of Earth's atmosphere, where the ionization is so complete that the particles form a plasma, a mixture of positively charged ions and negative electrons, that is constrained by Earth's magnetic field.

The magnetosphere forms the absolute limit of the planet – the hull of 'Spaceship Earth' (see Figure 6.2). On the upstream side of the solar wind, it deflects charged particles outwards past our planet. The interaction between the solar wind and Earth's magnetic field produces a shock wave at a distance of about 14 Earth radii, but the magnetosphere itself extends only to a distance of about 60,000 kilometres. Beyond this boundary, the magnetopause, lies interplanetary space.

Closer in, two doughnut-shaped zones of high particle density are centred above the equator at heights of 3000 and 15,000 kilometres; these are the radiation belts, named after James Van Allen, the American physicist who discovered them in the 1950s.

In the 1970s a later generation of satellites mapped the magnetosheath, a long tail of plasma streaming away downstream in the solar wind. In shielding Earth from the charged particles of the solar wind, the magnetic field deflects them into the Van Allen belts, but some spill over into the polar regions of the upper atmosphere, the polar cusps, where fast-moving electrons from the solar wind interact with atmospheric atoms to produce the dancing lights of the polar auroras.

Climate control for the planet

Earth's climate is remarkably stable, and has remained in a livable range for around 4 billion years, although, as we've seen, there have been a few worrying moments along the way. The key to its stability appears to lie in the interplay between plate tectonics and carbon dioxide in the atmosphere and oceans.

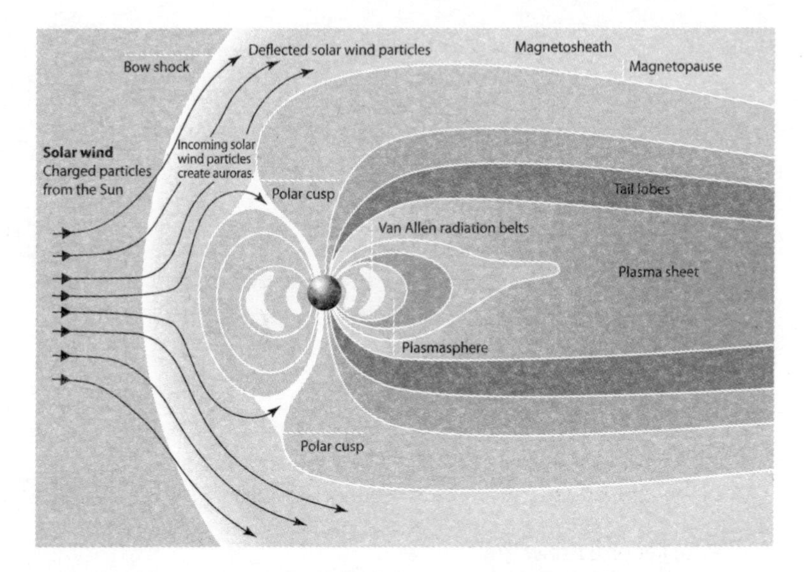

FIGURE 6.2 Earth's outer shield, the magnetosphere, deflects the solar wind and gives us the polar auroras.

The cycle begins with volcanoes spewing CO_2 into the atmosphere, which helps keep the planet warm. This warmth encourages seawater to evaporate, forming clouds and rain. As the rain contains dissolved CO_2, it is slightly acidic and so reacts with surface rocks in a process called chemical weathering – it dissolves carbon-containing minerals into the water. This mixture is then washed out to sea, where the minerals build up and eventually precipitate out to form new carbon-containing rocks on the seabed – sedimentary rocks (see Figure 6.3).

Sooner or later, plate tectonics carries these rocks into a subduction zone, where CO_2 is baked out of them by the heat of the mantle and later returns to the atmosphere via volcanoes.

This cycle turns out to be an extremely effective thermostat. When the planet warms, evaporation and rainfall increase, speeding the removal of CO_2 from the atmosphere and cooling

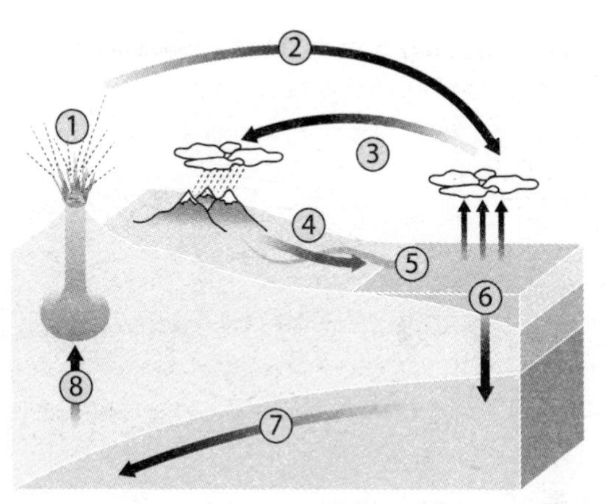

1. Volcanoes spew CO_2 into the atmosphere.
2. CO_2 keeps Earth warm via the greenhouse effect.
3. Warmth helps seawater evaporate, forming rain.
4. Rain contains CO_2 so is slightly acidic and dissolves minerals from the rocks into the water.
5. Dissolved carbon-containing minerals wash into rivers and into the sea.
6. Minerals precipitate out to form new carbon-containing rocks.
7. Rocks are eventually subducted into the mantle, where the CO_2 is released.
8. CO_2 returns to the atmosphere through volcanoes.

FIGURE 6.3 Carbon dioxide is at the heart of Earth's complicated climate-control mechanism.

the planet. When it is cold, rainfall decreases, allowing volcanic gases to build up in the atmosphere, which warms the planet.

On Earth, the Moon has also played a role in keeping the climate habitable. It dampens wobbles that would otherwise cause Earth's axis to tilt wildly: even small wobbles are enough to launch ice ages. Now, of course, humans are playing their part (see Chapter 10). The changes we make to the climate by burning fossil fuels could last millions of years but, after we've gone, Earth's underlying thermostat should be able to regain control. That, however, is not guaranteed.

The weather machine

If the atmosphere has been relatively well behaved over the long term, creating a fairly stable climate, the same cannot be said of the short term: our weather is prone to moods of spectacular beauty and stunning violence. These moods stem from the complicated interplay of the forces that set the troposphere in motion.

As the Sun warms Earth, the planet warms the parcel of air above it. In response, molecules in the parcel zing around more rapidly: its volume increases. Since this makes the air parcel less dense than the surrounding air, it becomes more buoyant and rises. Cooler, heavier air flows into the space it has vacated, where it in turn becomes heated and rises, continuing the cycle. This vertical movement of heat is called convection, and the rising parcels of air are known as thermals.

In this way, temperature differences cause variations in density and pressure that drive winds both vertically and horizontally as the air flows to try to equalize the pressure.

Earth's atmosphere is heated unequally at the poles and equator. This occurs because of simple geometry. We live on a sphere orbiting the sun, and sunlight falls from directly overhead on the equator, but at a sharply slanted angle near the north and south poles. The polar regions thus receive less energy from sunlight for a given area than the equator. This difference is the fundamental driver of weather on the planet. Heat naturally moves from hotter to colder areas, so the atmosphere and oceans transport heat from the equator to the poles. A planet without temperature differences would be a planet where the wind never blows. But on Earth the wind always blows, and sometimes it blows very hard.

Wind belts and the Coriolis effect

If Earth did not rotate, global wind patterns would be very simple. Hot air would rise at the equator, then spread out horizontally towards the poles once it reached the top of the atmosphere. At the poles it would cool, sinking as it became denser, then flow along the surface back to the equator. Surface winds would thus flow only from north to south in the northern hemisphere, and south to north in the southern hemisphere.

On a rotating sphere, the surface – and the air above it – moves fastest at the equator and not at all at the poles. Thus, Earth's easterly rotation deflects winds to the right, relative to their direction of motion, in the northern hemisphere, and left in the southern. This deflection is called the Coriolis effect.

Earth's rotation creates a Coriolis effect strong enough to produce three interlocking bands of surface winds in each hemisphere: the equatorial trade winds, the mid-latitude westerlies and the polar easterlies (see Figure 6.4). If Earth spun faster, there would be more of these wind belts. Jupiter's very fast rotation rate – its days last just ten hours – gives that planet many more bands of winds than Earth.

At high altitudes, fast west-to-east bands of wind called jet streams develop above the slower-moving surface winds. While this general pattern of wind belts predominates, because we do not live on a uniform sphere but on one with oceans, mountains, forests and deserts, actual wind patterns are far more complicated and variable.

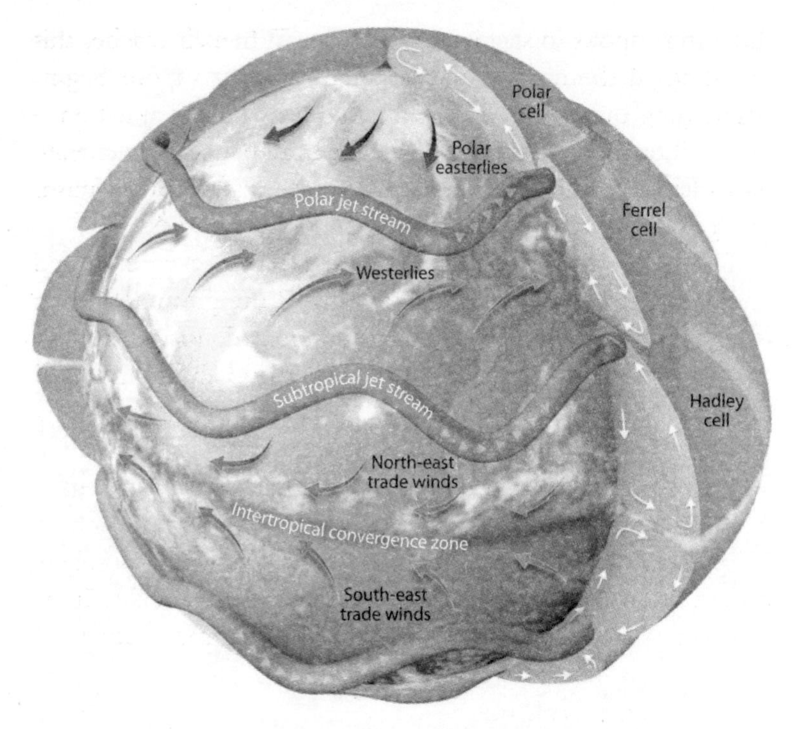

FIGURE 6.4 Earth's winds appear as distinct bands, which are created in part by the Coriolis effect.

Instability and precipitation

One of the prerequisites for severe weather is atmospheric instability, where a less dense volume of air rises above the surrounding air. Instability is greatest when there is cold, dense air aloft and warm, moist air below. The lower layer rises because warmth and water vapour both reduce air density. And in an unstable atmosphere, air that starts moving upwards keeps on moving upwards.

As the air rises, it expands further. Expansion causes cooling because as molecules of air move farther apart the kinetic energy in a given volume falls. At some point, it becomes too cool for the water vapour to stay in the gas phase. When air reaches this point, called the dew-point temperature, water vapour begins condensing out of the air, forming clouds and precipitation – rain, hail or snow. Thus, two ingredients are needed to generate precipitation: sufficient water vapour in the air and a mechanism to lift air so that it cools to its dew-point temperature.

The three main ways air gets lifted to cause cooling and condensation are through solar heating of the ground, causing thermals to develop; air masses of different densities meeting and creating 'fronts' that push air upwards; and air being forced upwards by mountains in its way.

While air temperatures fall with altitude in the troposphere, at an altitude of around 11 kilometres the air begins to warm again. This 'temperature inversion' marks the bottom of the stratosphere. No clouds form in the stratosphere, since air from the troposphere cannot rise above the inversion. This puts a lid on instability. If the inversion were missing, we would see more extreme weather.

Mid-latitude cyclones

The word 'cyclone' can be used to describe any rotating storm system and includes the ordinary low-pressure systems that develop in the mid-latitudes.

The unequal heating of the equator and poles often leads to storms thousands of kilometres across, which transport heat towards the poles. These are mid-latitude cyclones familiar for giving the mid-latitudes much of their precipitation.

Mid-latitude cyclones form where sharp temperature differences exist along a front separating cold, dry polar air from warm, moist tropical air. These great storms are primarily powered by the release of potential energy, as cold, dense air moving down and towards the equator displaces warmer, less dense air moving upwards and polewards.

An additional energy source is latent heat. A lot of energy is needed to turn liquid water into vapour, and this energy is released when the vapour condenses. When air is lifted and cooled in a storm, and water vapour condenses, the release of latent heat warms the surrounding atmosphere. That makes this air rise higher still, which releases yet more latent heat, powering the storm. The storm acts as as heat engine, converting heat into kinetic energy – wind.

Thunderstorms

When the Sun's heat is strong enough, the upward-moving thermals create puffy-topped cumulus clouds. In some cases, the tops of these cauliflower-shaped clouds may reach the top of the troposphere. In the upper parts of these clouds, freezing temperatures create ice and snow, and collisions between these particles separate electric charge. When the charge difference builds up to a critical level, a lightning bolt strikes, reuniting the positive and negative charges.

The cumulus cloud is now a cumulonimbus cloud – a thunderstorm. As well as providing life-sustaining rains for most of the planet, thunderstorms also bring hazards. The world's heaviest rainfall events are invariably caused by thunderstorms; the heavy rainfall of tropical cyclones is due to the thunderstorms embedded within them. Severe thunderstorms can generate destructive straight-line winds with speeds up to 240 kilometres per hour and generate hailstones as large as grapefruits. Finally, thunderstorms spawn nature's most violent windstorm – the tornado.

Tornadoes

During a tornado in Bridge Creek, Oklahoma, on 3 May 1999, Doppler radar measured a wind speed of 486 kilometres per hour about 30 metres above the ground – the fastest ever recorded. Winds of this strength cause total destruction, sweeping strong timber-frame houses off their foundations and badly damaging steel-reinforced concrete structures.

Since 2000 there have been nine top-end tornadoes in North America earning the official EF-5 designation on the Enhanced Fujita scale (winds exceeding 322 kilometres per hour). Unlike hurricanes, tornadoes are quite small, ranging from 75 metres across to about 3 kilometres. They descend from cumulonimbus clouds, which can be over land or water. Those that form or move over water are called waterspouts and tend to be much weaker than tornadoes that develop over land.

A very particular set of conditions is needed for tornadoes to form. Most important is the presence of instability and wind shear. A low-altitude flow of warm, moist air from an ocean area combined with a flow of cold, dry polar air high up creates the conditions for maximum instability, which means that parcels of air heated near the surface rise rapidly, creating powerful updrafts.

If a strong jet stream is present, with high winds near the top of the troposphere, there will be vertical wind speed shear. If the winds also change from southerly near the surface to westerly aloft, there is vertical wind direction shear. These two types of shear make the updraft rotate, creating a rotating thunderstorm, or supercell. Supercells spawn the vast majority of strong (EF-2 and EF-3) and violent (EF-4 and EF-5) tornadoes.

A third ingredient that is usually needed to generate supercell thunderstorms is the 'cap'. This is a region in the middle

FIGURE 6.5 Tornadoes are smaller than hurricanes but their violent winds do extensive damage.

layers of the atmosphere where dry, stable air has intruded. It prevents air rising very high until later in the day, when solar heating eventually generates enough instability for one thermal to burst through the cap. The result is a single, large supercell instead of a number of smaller, spread-out thunderstorms.

These conditions are most common in the Midwestern USA. The Gulf of Mexico provides a source of warm, moist air at low levels, and when this low-density air slides underneath high-density cold, dry air flowing southwards from Canada, an explosively unstable atmosphere often results. Add to this mix a mid-level intrusion of dry, stable air from the desert regions to the west and a powerful jet stream aloft creating plenty of wind shear, and dozens or even hundreds of tornadoes can result. During the tornado outbreak of 25–28 April 2011,

355 tornadoes – including four top-end EF-5s – ripped through 21 US states and Canada, killing 324 people.

While the vast majority of the world's tornadoes occur in the USA, they do affect other nations, too. Bangladesh averages three tornadoes a year, and many of these are strong and violent. The world's deadliest tornado was a violent twister that hit Bangladesh on 26 April 1989, killing more than 1300 people.

Tropical cyclones

Hurricanes, typhoons, tropical storms and tropical depressions are all examples of tropical cyclones. They form only over warm ocean waters of at least 26 °C and, unlike storm systems over land, derive their energy exclusively from the latent heat released by condensing water vapour.

Like tornadoes, tropical cyclones need a particular set of ingredients in order to form, starting with warm ocean water. In addition, vertical wind shear must be very low: in other words, the difference in wind speed between the surface and the top of the troposphere must be less than around 10 metres per second. Any faster and the shear tilts and stretches the core of a developing tropical cyclone, carrying away its heat and moisture.

Strong upper-level winds associated with the jet stream or upper-level low-pressure systems are the most common source of wind shear. The subtropical jet stream tends to be closer to the equator in winter and spring, which is why hurricanes and typhoons rarely form in the Caribbean Sea or western Pacific in those seasons, even though ocean temperatures are warm enough year-round to support such storms. Tropical cyclones also need high moisture through a deep layer of the atmosphere. Dry air from Africa or North America often disrupts a hurricane in the process of forming.

Finally, a tropical cyclone needs something to get it spinning. In the Atlantic this is usually provided by disorganized areas of low pressure called African easterly waves, which emerge from the coast of Africa and move westwards towards the Caribbean. Hurricanes get still more spin from the effect of Earth's rotation. The Coriolis effect is zero at the equator and maximum at the poles, so tropical cyclones generally cannot form within about 5 degrees of latitude from the equator. But once going, they tend to expand as they move polewards, thanks to the increasing vertical spin from the Coriolis effect.

Typhoon or hurricane?

A tropical cyclone starts life as a tropical depression – an organized, spinning storm system with wind speeds of less than 63 kilometres per hour. When the winds grow faster, the system is given a name and classified as a tropical storm. When the winds reach 119 kilometres an hour, a ring of intense thunderstorms called the 'eyewall' forms around the storm's centre. Within the eyewall is the 'eye' of the storm, a clear, calm region of sinking air.

Once the sustained winds exceed 119 kilometres an hour, the storm is classified as a hurricane if it is in the Atlantic or eastern Pacific, and as a typhoon in the western Pacific. In the Indian Ocean or in the southern hemisphere it is simply called a cyclone or tropical cyclone. There is no meteorological difference between these differently named storms.

Very warm water extending to a depth of 50 metres or more can help fuel rapid intensification of a tropical cyclone to 'major' hurricane status, with winds of 178 kilometres an hour or more – the most fearsome and destructive type of storm on the planet.

Traditionally, hurricanes are ranked from 1 to 5 on the Saffir–Simpson scale, based on the maximum sustained wind speed. However, this scale can be misleading. A weak storm that covers a huge area of the sea can generate a larger storm surge than a smaller but more intense hurricane with a higher Saffir–Simpson rating. To give a better idea of storm surge potential, the experimental Integrated Kinetic Energy scale has been developed. It is a measure of both wind speed and the area over which high winds extend.

Monsoon depressions

Monsoons operate via the same principle as the familiar summer afternoon sea breeze, but on a grand scale. In summer, the land gets hotter than the sea: that's because on land, the Sun's heat is concentrated close to the surface, while at sea, wind and turbulence mix warm water at the surface with cooler water lower down. Also, it takes more energy to raise the temperature of water than it does to heat the soil and rock of dry land.

As a result, a low-pressure region of rising air develops over land areas. Moisture-laden ocean winds blow towards this region and are drawn upwards when they reach land. The rising air expands and cools, releasing its moisture as some of the heaviest rains on Earth – the monsoons.

Each summer, monsoons affect all continents except Antarctica and are responsible for rains that sustain the lives of billions of people. In India, home to 1.34 billion people, the monsoon provides 80 per cent of the annual rainfall. Monsoons have their dark side, too: hundreds of people in India and surrounding nations die every year in floods and landslides triggered by the heavy rains.

The most deadly flooding events usually come from monsoon depressions, also known as monsoon lows. A monsoon depression is similar to, but larger than, a tropical depression. Both are spinning storms hundreds of kilometres in diameter with sustained winds of 50 to 55 kilometres per hour, nearly calm winds at their centre and very heavy rains. Each summer, some seven monsoon depressions form over the Bay of Bengal and track westwards across India. In 2010, two major monsoon depressions crossed India into Pakistan in July and August, bringing heavy rains and causing the most costly floods in Pakistan's history ($10 billion).

The riddle of rain?

While we know plenty about how weather works, there are lots of questions we cannot yet answer. Weather is, as we all know, chaotic, which makes forecasting tough. We also struggle to understand events that are sporadic, short-lived and take place in confined locations high above our heads. The next two sections, and another in Chapter 10, are examples of this latter category. They focus on research into an area that is gaining increasing attention: what's really going on in clouds.

Clouds are both familiar and mysterious. They form when water vapour condenses into minute droplets or ice crystals. Yet when it comes to releasing that water, we're still in the dark as to why some clouds unleash a torrent while others don't shed a single drop.

The riddle stems from the physics of ice formation. Clouds produce rain or snow when the droplets they contain grow big enough to overcome atmospheric updraughts. Most of the time, falling requires freezing – ice crystals grow faster than droplets, meaning that they reach falling weight before they are swept away to evaporate and vanish. But strange though it may seem, pure water in the atmosphere can remain liquid down to −40 °C. And although the molecular secrets behind this phenomenon are still puzzling, it means that water droplets in clouds usually need a bit of help to form ice.

FIGURE 6.6 Clouds often need something in the air to help them generate a deluge.

That help comes in the form of 'ice nucleators', airborne particles or aerosols that provide a tiny object around which water molecules arrange themselves into the lattice structure of an ice crystal. Salts thrown up from ocean spray and mineral dust from desert winds can do the trick, and they are abundant in the skies. But they cannot seed ice crystals above −15 °C, which is the temperature inside up to half of all clouds that form over land. There must be something else lurking in these common clouds.

But what? One potential ice-maker's identity emerged in the early 1970s, when researchers showed that a leaf-dwelling bacterium called *Pseudomonas syringae* is a catalyst for ice formation even in relatively warm conditions. Why *P. syringae* evolved these instant-freeze powers isn't clear, but it might have been a way to get into a plant's tissues: spiky ice crystals pierce leaves and rip open cells, serving up the nutrients inside.

In spring 1978 a plant pathologist named David Sands of Montana State University in Bozeman hired a small plane and discovered *P. syringae* floating in clouds high above the ground. He proposed that drizzle and downpours were summoned by bacteria, an idea that did not go down well with atmospheric scientists.

A decade or so after that discovery, researchers managed to isolate one of the genes that make ice-nucleating proteins. Many more species of microorganism boasting this ability then came to light, including various species of fungi. But still no one took Sands's idea seriously.

That began to change in 2007, when scientists collected fresh snow from around the world and looked at the ice-nucleators. They placed them in pure water and cooled them to identify samples that froze at temperatures above −7 °C. These they then heated to destroy any proteins – on the assumption that this would deactivate any biological nucleators. When the droplets cooled again, most no longer froze above −7 °C, indicating that most of the ice-nucleating particles were biological. Then, in 2015, similar research into giant hailstones found that they too were born when biological particles transformed water into ice.

And there are plenty of them up there. In recent years, we have found all kinds of microorganisms living at altitudes where they could influence the workings of clouds. One study collected samples at up to 10,000 metres above the Atlantic Ocean, Caribbean Sea, Gulf of Mexico and the continental USA as hurricanes Earl and Karl passed through. It clocked 314 different species of bacteria, most of which were alive. It also found roughly as many biological cells as soil and dust particles.

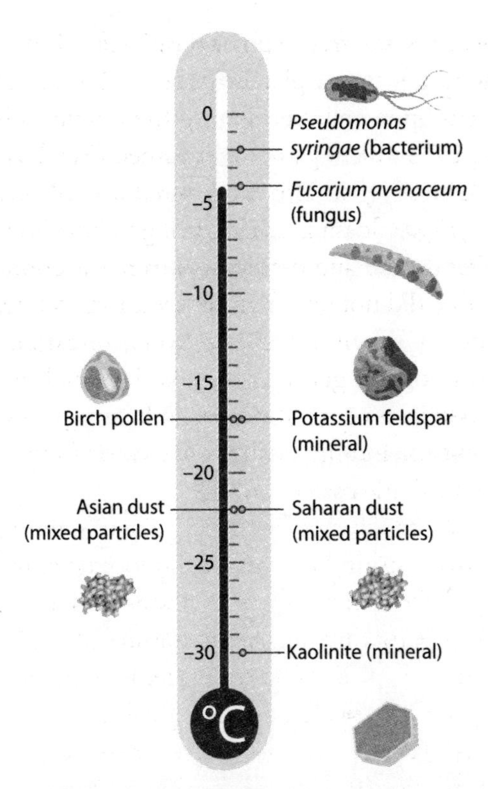

FIGURE 6.7 To make it rain, clouds often have to form ice crystals. Pure water can stay liquid down to −40 °C but microscopic particles can kick-start ice formation at higher temperatures.

That doesn't mean that these cells are affecting clouds or making rain, however. To find that out, you need to see what they're up to inside clouds. Which is exactly what Kim Prather, an atmospheric chemist at the University of California, San Diego, has done by sucking in ice crystals from rain clouds. Prather and her colleagues found that roughly 40 per cent of the particles on which ice crystals had formed in the most rain-laden clouds are biological in origin.

This amounts to tantalizing evidence that microbes do indeed seed ice in warm clouds. But Prather has yet to catch bacteria in the act. Even if she did, some people question whether there are sufficient numbers up there to make a difference. Soot and natural mineral particles are thought to be more abundant, and therefore more likely to hold sway. Nevertheless, advocates propose that microbes may yet have significant effects in certain regions and times of the year.

What if microbes make it rain?

If Sands and other supporters of the idea of living ice-nucleators are correct, it raises a host of questions and possibilities. For example, did *P. syringae* evolve to make use of trips to the clouds to ensure its own dispersal – and to ensure a good drenching for its plant hosts, too?

And how exactly do microbes affect weather patterns? Sands points to evidence that some types of vegetation release more ice-nucleating bacteria after rainfall, generating further drenchings. In turn, this raises the question of whether we have been unwittingly modifying the weather by practising agriculture.

If we have accidently shaped weather in the past, can we do it deliberately in the future? The USA has a long tradition of sending up planes to catalyse ice formation by spraying the skies with silver iodide. In 2015 Los Angeles County responded to drought in California by paying out half a million dollars to a cloud-seeding contractor, attracting criticism because there is scant evidence that silver iodide increases rainfall.

Sands thinks natural ice-makers could offer a better solution: identify or create plants that harbour ice-nucleating bacteria, sow them in the right places and let them seed rain, for example.

For now, all this is speculative. And there's one other issue that needs to be taken into account. You can't induce rainfall where there is no water vapour in the air, and the amount of moisture available to fall in a given place will always be dependent on global weather patterns – and they are changing fast.

Bolt from the blue

Killer storms, such as hurricanes and tornadoes, are thankfully relatively rare. Of the more common weather hazards, there's one you really don't want to get up close and personal with: lightning.

It's a very common phenomenon – roughly a hundred lightning flashes happen somewhere around the globe every second. But because of the difficulties and dangers that go with studying lightning, we still know comparatively little about how a flash begins. Only slowly are researchers shedding new light on the conditions that create this spectacle. And it looks as though we may have had lightning back to front all along.

We've all probably experienced some degree of pain from a spark, usually after shuffling across a shaggy carpet and reaching for a metal doorknob. As you walk across the carpet, friction between your feet and the floor scrapes negatively charged electrons from atoms in the carpet, which run up through your body and give you an overall negative charge. This build-up of charge may seem trivial, but over short distances the electric field it generates can grow surprisingly large.

As your negatively charged finger edges forwards it repels electrons in the doorknob, leaving it positively charged nearest your hand. These separated charges generate an electric field which, if it reaches the critical value of 3 million volts per metre, causes the air between hand and knob to break down. Electrons are ripped from molecules in the air, turning it from insulator to conductor. In a split second, your surplus electrons flow painfully out of your finger and through the ionized air to the knob. Ouch! Electrical neutrality is restored.

Friction is the root of the electric field in thunderstorms, too, although its origin is different. In thunderclouds, powerful updraughts carry ice crystals to their tops while heavier hail falls towards Earth. Friction between these two streams robs the ice crystals of electrons, so the upper cloud becomes positively charged and the lower part negatively charged. This creates an electric field like the one between you and the doorknob, but when the field in a thundercloud grows strong enough to ionize air, the results are rather more spectacular. Electrons carve ionized channels, or leaders, through the air, hunting for the nearest positive charge.

During a familiar cloud-to-ground lightning flash, the negative charge finds this at Earth's surface. Yet the most common

FIGURE 6.8 Research into how lightning flashes begin is generating counter-intuitive results.

type of lightning is an intra-cloud flash, where the discharge runs up to the positive region at the top of the cloud. Either way, once one of these leaders reaches a region of opposite charge, electric current shoots between the two points to generate lightning flashes five times as hot as the surface of the sun.

That's been the majority view. But it has a problem: despite having sent balloons and aircraft laden with instruments into thunderclouds since the 1950s, we have never measured that 3-million-volts-per-metre electric field needed to break down air. Instead, the field is typically a tenth of this value, suggesting that lightning operates in a different way from a conventional electric spark.

Extraterrestrial assistance

One potential solution to this problem comes from a surprising source: outer space. Every second, billions of high-energy particles crash into our atmosphere. If one of these were to collide with an electron in a thunderstorm, the electron would receive a serious injection of speed. As it ripped through the cloud, this runaway electron would ionize large numbers of air molecules, generating an avalanche of other high-energy electrons.

The thinking goes that the resulting sudden accumulation of charge briefly intensifies the local electric field. Although the details are still sketchy, the added effect of this field may be enough to spark lightning, in an effect known as a 'runaway breakdown', without the background electric field being anywhere near 3 million volts per metre.

Support for this idea came in 1991, shortly after NASA launched the Compton Gamma Ray Observatory into orbit. Its mission was to search for gamma rays, the most powerful form of radiation in the universe, usually created when stars explode. So it was a surprise when the observatory detected these high-energy photons coming not from distant galaxies but from thunderstorms in Earth's atmosphere.

Physicists were quick to connect the dots. As cosmic-ray-accelerated electrons zigzag their way between collisions with

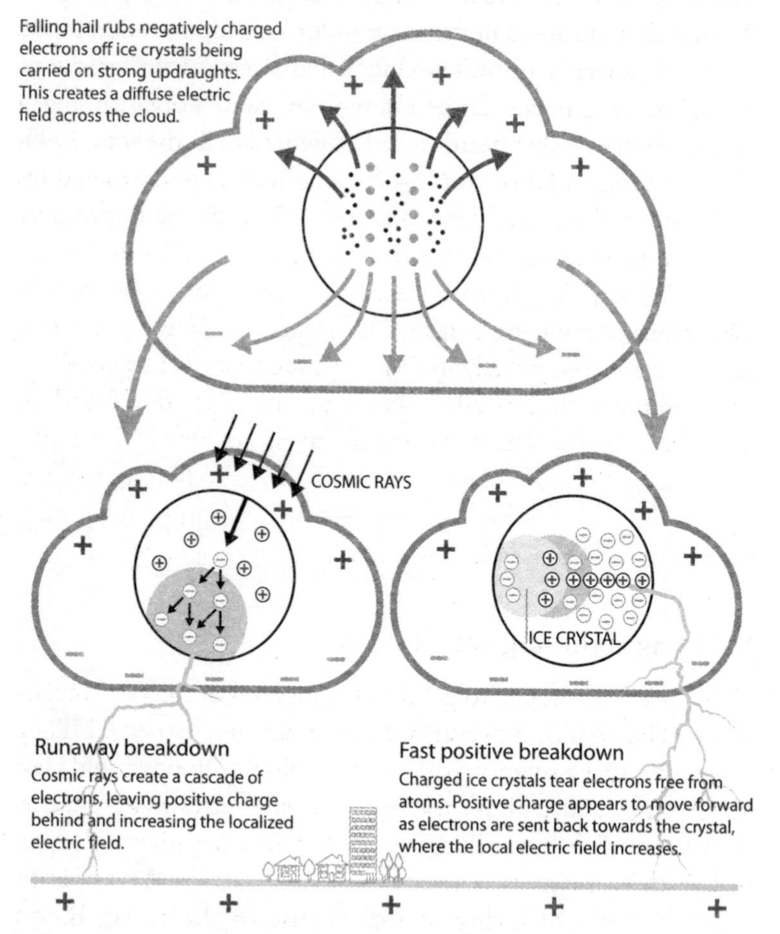

FIGURE 6.9 The electric fields inside thunderstorms appear to be too low to cause lightning on their own. Two processes may boost local fields enough to trigger a flash.

air molecules, they produce not only further high-energy electrons but also high-energy photons. Gamma rays are a sign that runaway breakdown is at work.

As enticing as this link appears to be, it could be a coincidence: we have no direct evidence that this is what's going on. To find that, we need better ways to look inside thunderclouds.

Luckily, there is a way. The fluctuating electric fields produced by lightning generate large amounts of radio noise, causing a crackle similar to that heard on analogue radios. In the mid-1990s physicist William Rison at New Mexico Tech in Socorro and his colleagues realized that they could use GPS receivers to precisely map that radio noise, and thus the lightning flashes. Today Rison's Lightning Mapping Array stretches across 16 stations in the nearby Magdalena Mountains. It takes 3D images of lightning within a thundercloud, but initially its time resolution was not so good.

To improve this, in 2016 Rison and his team developed an interferometer to detect the radio waves and attached a high-speed camera capable of taking more than 180 million frames per second. With all this kit, they get video images and a full, three-dimensional map of lightning bolts – the most accurate yet.

Standing lightning on its head

The researchers hoped to get detailed pictures of runaway breakdown. The results, however, weren't what they expected. They discovered that a mighty spark deep within the clouds could raise the electric field without the need for extraterrestrial assistance. And rather than beginning, as expected, in a negatively charged region and running to a nearby positive area, it did the opposite.

After weeks of trying to make sense of the result, Rison and his team realized that the perceived flow of positive charge amounted to a rush of electrons moving in the opposite direction. The culprit, in their view, might well be a polarized ice crystal with a negative charge on one side and a positive charge on the other. If the positive charge grows strong enough to tear electrons from nearby air molecules, it can create a further positive charge. This then behaves like the crystal's new tip, creating yet another positive charge, a process that repeats itself for as long as the electric field is strong enough to sustain it.

They called this process 'fast positive breakdown'. It creates a little wiggly ribbon of ionized air that starts growing past the end of the ice crystal. As that positive 'streamer' grows, it's like a vacuum cleaner sucking up negative charge. And it sends that charge back towards the ice crystal, like a swimmer pushing water behind her. Once the ice crystal accumulates enough added charge, it can create a lightning leader, the forerunner to a strike.

Astonishingly, Rison's results suggest that potentially all flashes seen in thunderstorms may be initiated by fast positive breakdown. Cosmic rays may still play a part, of course. But some in the field see Rison's evidence as a game-changer, like finding the corner of a jigsaw that gets you started.

Spark of recognition

From flashes of light that resemble sea monsters to electric orbs thought capable of melting glass windows, lightning comes in bizarre forms:

Sprites: Once thought to be a myth, sprites are fleeting flashes of red light high above the clouds that look like giant jellyfish. These are believed to be produced by the strong electric fields generated in the upper atmosphere when lightning hits the ground. We don't yet understand exactly how they form.

Elves: These glowing doughnut-shapes of light grow to 400 kilometres across and then disappear – all in less than a millisecond. They are thought to arise when the electric field in a cloud causes electrons to smash into nitrogen molecules, which in turn give off a distinctive red glow.

Blue jets: Occasionally glimpsed from the International Space Station, blue jets snake upwards from thunderclouds to a height of roughly 50 kilometres and then vanish one-tenth of a second after they begin. Their comparative rarity has made it difficult to understand their origins.

Upside-down lightning: When it comes to neutralizing the electric field between a thundercloud and the planet's surface, why should the cloud always have to give way? Sometimes lightning forms at ground level and shoots upwards until it hits the cloud. The specifics of when and how it strikes remain a mystery.

Ball lightning: Orbs alive with electricity have been seen melting glass windows, floating through buildings and even bouncing down the aisles of aircraft. Although people have reported ball lightning in nature for more than 2000 years, scientists are unsure what is really going on. There are many theories, but none is fully accepted.

7
Oceans

'How inappropriate', wrote Arthur C. Clarke, 'to call this planet Earth when clearly it's Ocean.' It's hard to argue with that: viewed from space, our planet looks predominantly blue because the oceans cover 71 per cent of the surface. We wouldn't be here if it were otherwise.

Earth's restless waters

Implicit in Arthur C. Clarke's comment is that we take the seas for granted. Worse, we abuse them. They are still a significant source of food for many around the world, yet overfishing has endangered 85 per cent of the world's fisheries. And just as with the atmosphere, we have tended to treat the oceans as a giant rubbish tip. Sadly, pollution from agriculture, industry and everyday living is still eroding the oceans' ability to support life (see Chapter 9).

On a planetary scale, oceans provide other 'services'. They act as enormous storage heaters, for example, smoothing out temperature extremes and giving Earth an equable climate for fostering life. In recent decades, another of the oceans' faculties has come to the fore: they consume vast amounts of carbon and isolate it from the atmosphere.

This ability rests in large part with the ocean circulation, a network of currents transporting huge quantities of heat, salt and other things around the world. The volume of water moving in any one of the world's five largest surface currents, such as the Gulf Stream, is some 50 times the combined flow of all Earth's freshwater rivers.

Water on the ocean surface absorbs CO_2 from the air, locking it away as it sinks to the ocean depths. So, too, do tiny marine plants called phytoplankton. If climate change disrupts the pattern of ocean circulation, as some scientists suggest, it could seriously limit the ocean's ability to soak up CO_2. This means that we can only fully understand climate change if we get to grips with the role of the oceans.

Driving forces

Ocean currents are driven by several factors. Winds above the ocean drive surface ocean currents. Another leading player is the Coriolis force, caused by Earth's rotation. The force gives a sideways kick to ocean currents, just as it does to winds (see Chapter 6).

In the northern hemisphere, the Coriolis force deflects ocean currents to the right, relative to their direction of motion, and in the southern hemisphere to the left. The basic pattern of circulation at the surface of the oceans ends up as a number of enclosed flow patterns called gyres. Because the Coriolis force increases from the equator to the poles, it turns out that the gyres have an asymmetrical shape. In the western margins of ocean basins, there are strong, narrow and fast gyre currents with speeds of up to 2 metres per second. These 'western boundary currents' include the Gulf Stream in the North Atlantic. In the east, gyre circulation consists of weak, broad and slow flows of less than 0.1 metres per second.

All the major ocean basins have gyres (see Figure 7.1). Subtropical gyres are large closed patterns of circulation found between about 10 and 40 degrees latitude. Subpolar gyres occur between about 50 and 70 degrees. The wind piles up the water within these gyres. This means that sea level is not flat. For example, in the North Atlantic subtropical gyre, the sea level of the Sargasso Sea – between the West Indies and the Azores – is about a metre higher than the sea around Bermuda. The major exception to these gyre patterns is the Antarctic Circumpolar Current, a strong current that flows around the Southern Ocean from west to east. There are also complex patterns of currents driven by winds along the equator, where the Coriolis force vanishes.

The fate of ocean water does not end here, however, because it depends in large part on an interplay between its temperature, its density and its salinity. Salinity varies throughout the world's oceans. Typically, 1000 grams of seawater contains 35 grams of

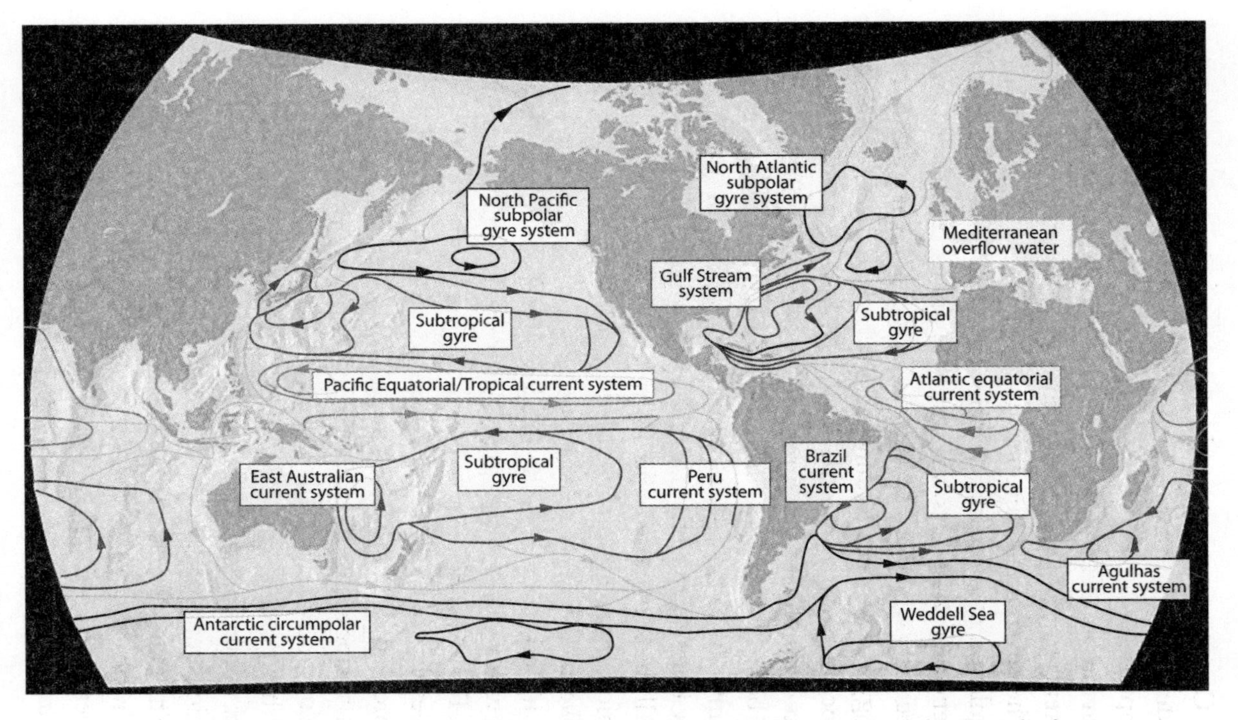

FIGURE 7.1 The major ocean surface currents are driven by the winds and the Coriolis force.

salt with, for example, lower values in the Baltic Sea and Arctic Ocean, and higher values in the Mediterranean and Red Seas. Water temperature also varies greatly around the globe. Both temperature and salinity help to determine the density of seawater, so density also varies from place to place and helps to drive ocean circulation.

The warmer ocean water gets, the less dense and lighter it becomes, causing it to rise towards the surface. The converse is true: the colder water is the denser it gets, causing it to sink. And the more salt there is in the water the denser it becomes. So the greatest contrast in density exists between cold, heavy saltwater and warm, light freshwater.

In winter in polar regions the ocean surface loses heat to the atmosphere, eventually leading to the formation of sea ice and increasingly salt-rich water. The now heavy, cold, salty polar water sinks to great depths, causing surface water to be dragged from tropical regions towards the poles to replace it. Meanwhile, in the depths of the polar seas, heavy water piles up and is pushed towards the equator. On this journey the water becomes warmer and lighter. Thus a huge oceanic loop is completed. The whole process of turning warm, fresh surface water into cold, salty deep water is called deep convection and acts as a kind of heat engine driving ocean circulation.

The pattern of flow driven by temperature and salt differences is called the thermohaline circulation, and is separate from the surface wind-driven circulation. The combined effect produces a three-dimensional circulation known simply as the oceanic conveyor belt (see Figure 7.2). The circulation of water in the Atlantic, Indian and Pacific ocean basins is linked by the Southern Ocean. The climate of Western Europe relies heavily on the conveyor belt. Thanks to the North Atlantic Drift – an extension of the Gulf Stream – Western Europe

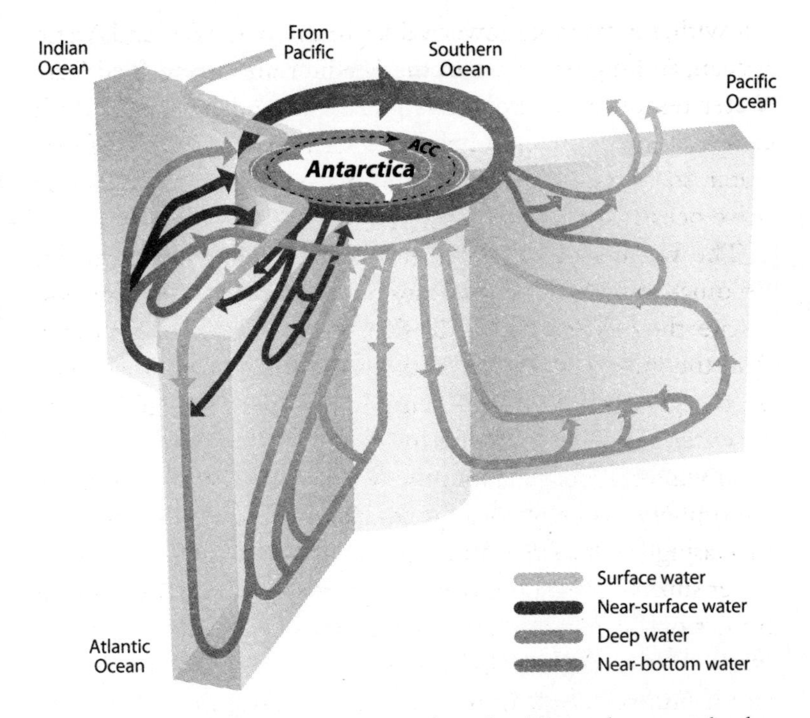

FIGURE 7.2 Cold, dense water near the poles sinks to the ocean depths where it flows towards the tropics. This sinking water draws warm water along the surface from the tropics, creating a 'conveyor belt'.

gets 'free' heat – equivalent to the output of about a million power stations.

Seawater can store a great deal of heat – the top 3 metres of the world's oceans has the same thermal capacity as the entire atmosphere. The Sun heats the ocean surface directly, so surface waters are usually the warmest and lightest, while water near the ocean bottom is the coldest and densest. This gradual layering or stratification of the oceans is important because water can move more freely along layers of constant density than it can across them. But water can still drop from one depth to another in regions where there is deep convection or, as will become clear, where ocean layers tilt into the depths.

Waters old and new

Surface winds churn up the water in a mixed layer, where temperature and density are roughly constant, which stretches down about 100 metres. Beneath this mixed layer, the temperature and density vary according to depth. Most of the ocean contains cold water that has sunk at high latitudes in winter. Because warm tropical waters are lighter, they float on top of the colder polar waters. The relatively sharp divide between them, where the temperature drops rapidly with depth, is known as the thermocline.

The existence of the thermocline depends crucially on the thermohaline circulation. Water sinks relatively rapidly in a few sites of deep convection – such as in the polar waters of the North Atlantic and in the Weddell Sea, near Antarctica. But it 'upwells' much more gradually, around a centimetre a day on average, across a huge area in the warmer temperate and tropical regions. It then returns slowly towards the pole near the surface, to complete the cycle.

Because surface polar waters are colder and denser than surface tropical waters, ocean density layers don't usually run parallel to the sea surface. Instead, layers of constant density – called isopycnals – generally slope downwards from the ocean's surface. This is why, outside the local areas of deep convection, water from the surface can still penetrate into the ocean. The water carries with it some of the heat, salt, oxygen, CO_2 and other characteristics it had at the surface. This ferrying of water properties into the ocean interior is known as 'ventilation'. It can isolate CO_2 dissolved in surface seawater from the atmosphere and it's also why changes in ocean circulation patterns affect the oceans' ability to absorb atmospheric CO_2.

How old is water?

Oceanographers say the age of ocean water is the time since that water was last at the surface of the ocean. Studying the age of water helps them to piece together where water has come from, and where it is going.

To age water, oceanographers use natural radiocarbon (carbon-14), dissolved oxygen and silica. They can also use pollution from human activity, such as tritium, the radioactive isotope of hydrogen left over from nuclear bomb tests in the 1950s and 1960s. Water near the surface may be only a few years old or less, while deeper water will be much older. For example, water in the North Atlantic colder than 2°C was probably last in contact with the atmosphere 200 years ago, while, water deeper than 1500 metres in the Pacific Ocean is probably more than 500 years old.

The vertical circulation of ocean water means that even very old water must inevitably return to the surface before starting to sink again. Water can travel great distances and take thousands of years to complete a single path of circulation. It may take many twists and turns. Just as winds on land are channelled through valleys and deflected by hills, so are deep ocean currents affected by the shape of the ocean floor – its bathymetry.

Narrow channels, where the flow is constrained by bathymetry, connect various parts of the world's oceans. At Gibraltar, for example, Atlantic water that has a relatively low salinity flows into the Mediterranean Sea, while salty Mediterranean water flows out deep down. Such bottlenecks may be crucial for regulating the exchange of waters with different properties in the global ocean circulation.

Ocean eddies

The picture, then, might seem to be one of a steady, slow-moving circulation. However, the fine detail is missing. Just as the atmosphere has its 'highs' (anticyclones) and 'lows' (cyclones), so the ocean has its eddies – ocean 'weather systems' that move at speeds of a few centimetres a second, carrying heat and salt from one part of the ocean to another. These eddies, though, are around 100 kilometres across – much smaller than their atmospheric counterparts which can stretch to 1000 kilometres.

In 'fast-moving' regions such as the Gulf Stream, eddies may last only a couple of months before being swallowed up. By contrast, eddies travelling in less dynamic regions may survive for two years or more. Eddies form as the result of unstable flow, such as where a strong current meanders, or where there is a large difference in speed between neighbouring parts of a current, or where there is a significant change of velocity with depth.

Eddies carry a lot of kinetic energy and heat and can play an important part in the energy and heat budgets of the oceans. Eddies may also be important in transferring water characteristics between oceans. The Agulhas Current – a western boundary current off the east coast of Africa – regularly generates eddies that carry warm, salty water from the Indian Ocean into the South Atlantic.

Today, unravelling the mysteries of ocean circulation is a huge task involving scientists from many countries. If we are successful, then we will have taken one more step towards understanding Earth's climate, and perhaps being able to predict its future.

The real monsters of the deep

Given the inherent difficulties of studying the deep oceans, it makes some sense that we are only now beginning to understand what goes on down there. More surprising, however, is that we are still learning about phenomena that take place in plain sight on the sea surface. A few decades ago the notion of one-off giant waves, that could seriously damage or sink large ships, was rejected as salty sea-dogs' tales. But not any more. Real-world observations, backed by improved theory and lab experiments, leave no doubt that monster waves happen – and not infrequently.

Science has been slow to catch up with these rogue waves. There is not even a universally accepted definition. One with wide currency is that a rogue is at least double the significant wave height, itself defined as the average height of the tallest third of waves in a region. What this amounts to is heavily dependent on context: on a calm sea with significant waves 10 centimetres high, a wave of 20 centimetres might be deemed a rogue.

FIGURE 7.3 How do we predict colossal waves like the one in this photo illustration?

If that seems puny, for a long time the oceanographers' models predicted that anomalous, tall waves barely existed. Those models rested on the principle of linear superposition: that when two trains of waves meet, the heights of the peaks and troughs at each point simply add together. It was only in the late 1960s that Thomas Brooke Benjamin and J. E. Feir of the University of Cambridge, UK, spotted an instability in the underlying mathematics. When long-wavelength waves catch up with shorter-wavelength ones, all the energy of a wave train can become abruptly concentrated in a few monster waves – or just one.

The pair went on to test the theory in experiments at a large wave tank at the National Physical Laboratory in south-west London. Near the wave-maker, which perturbed the water at varying speeds, the waves were uniform and civil. But about 60 metres on, they became distorted, forming into short-lived, larger waves that we would now call rogues.

It took a while for this new intelligence to trickle through. Theory and observation finally came together in 1995 in the North Sea, about 150 kilometres off the coast of Norway. New Year's Day that year was tumultuous around the Draupner sea platform, with a significant wave height of 12 metres. At around 3.20 p.m., however, accelerometers and strain sensors on the platform registered a single wave towering 26 metres over its surrounding troughs. That's half the height of Nelson's Column. According to the prevailing wisdom, this was a once-in-10,000-year occurrence.

The Draupner wave ushered in a new era of rogue-wave science. In 2000 the European Union initiated the MaxWave project. During a three-week stretch early in 2003, it used boat-based radar and satellite data to scan the world's oceans for giant waves, and turned up ten that were 25 metres or more in height.

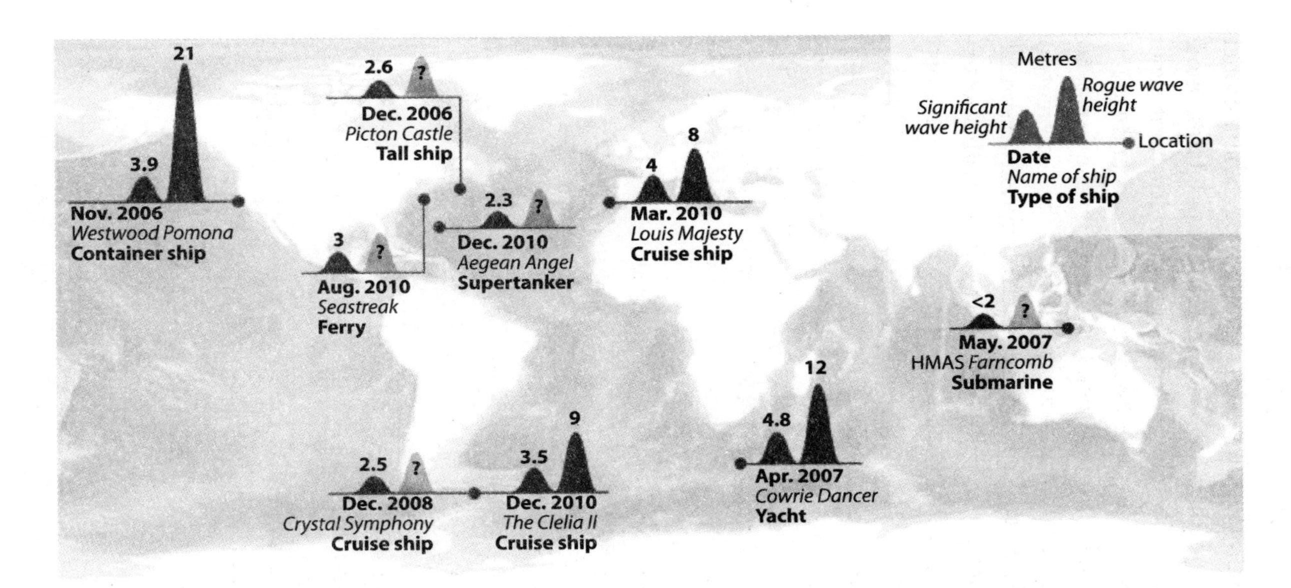

FIGURE 7.4 Nine verifiable incidents of ships hit by rogue waves in deep-ocean waters were drawn from an analysis of media reports between 2006 and 2010.

We now know that rogue waves can arise in every ocean. The North Atlantic, the Drake Passage between Antarctica and the tip of South America, and the waters off the southern coast of South Africa, are particularly prone (see Figure 7.4). This all casts historical accounts in a new light and rogue waves are thought to have played a part in the unexplained losses of some 200 cargo vessels in the two decades to 2004. In 2014 a freak wave struck the cruise ship *Marco Polo* in the English Channel, smashing windows and killing a passenger.

Making a monster

What makes a rogue wave in the real world? By combining theory and observation in increasingly complex computer models, researchers are building up a catalogue of real-world conditions that generate rogues. One is when a storm swell runs into a powerful current flowing in the opposite direction. This is often the case along the Gulf Stream in the North Atlantic, and where sea swells run counter to the Agulhas current off south-eastern Africa. Another is a 'crossing sea', in which two wave systems – often one generated by local winds and a sea swell from further afield – converge from different directions and create instabilities.

Louis Majesty

When the cruise ship *Louis Majesty* left Barcelona in Spain for Genoa in Italy, it was for the final leisurely leg of a tour. But the Mediterranean Sea had other ideas.

Storm clouds were gathering as the boat left port at around 1 p.m. on 3 March 2010. The sea swell steadily increased during the first hours of the voyage, though it was never out of the ordinary. Then, at 4.20 p.m., without warning, the ship ran into a wall of water 8 metres or more in height. As far as events can be reconstructed, the boat's pitch as it descended the wave's lee tilted it into a second, and possibly a third, monster wave immediately behind.

Water smashed through the windows on deck 5, almost 17 metres above the ship's water line. Two passengers were killed instantly and 14 more injured. As suddenly as they had appeared, the waves disappeared and the ship limped back to Barcelona. Today researchers think the incident may have been caused by a crossing sea. When they fed wind and wave data into a model to 'hindcast' the state of the sea in that area at the time, it indicated that two wave trains were converging on the ship, one from the north-east and one from the south-east.

Simpler situations might generate rogues, too. In December 1980 a cargo carrier loaded with coal lost its entire bow to an estimated 20-metre wave in the 'Dragon's Triangle', a region of the Pacific south of Japan notorious for accidents. A Japanese investigation had blamed a crossing sea, but when researchers used a more sophisticated wave model to hindcast the conditions, they found it likely that a strong gale had poured energy into a single wave system far larger than conventional models would predict.

Models of such single-system wave propagation have changed at some pace, and so have their predictions. In 2012 researchers found these models even permit the possibility of 'super rogues' towering as much as 11 times the height of the surrounding seas, a possibility since borne out in wave-tank experiments.

With growing appreciation of rogues, plans are now being laid to warn sailors when they are heading into conditions that might create them. There is still some way to go to perfect such systems, but at least we have the theory and models in place to forecast phenomena whose existence was doubted even up to a few decades ago.

Cryptic rivers

Beneath Turkey's Bosphorus Strait flows a mysterious river. It has banks and rapids and in places is a kilometre across. If it snaked across the land, the size of its flow would make it our sixth largest river. Yet this current flows silently 70 metres beneath the sea surface from the Sea of Marmara into the Black Sea.

The hidden river has no name, and is by no means unique. Myriad underwater rivers criss-cross the ocean floor, some many thousands of kilometres long, tens of kilometres wide

FIGURE 7.5 A current flowing beneath the Bosphorus Strait in Turkey has started to reveal the secrets of undersea rivers.

and hundreds of metres deep. They shunt sediments into the oceans, carrying with them the oxygen and nutrients that enable life to thrive at great depths.

Drain all Earth's oceans and you would find that underwater rivers have gouged a maze of conduits known as abyssal channels. To look at, they resemble terrestrial rivers, but their flows can behave much more like avalanches, dust storms or pyroclastic flows from volcanoes. Their destructive power makes them a major hazard for the telecommunication cables that snake along the ocean floor.

In fact, the existence of these channels was first deduced from the snapping of cables. On 18 November 1929, a 7.2-magnitude earthquake struck the Grand Banks, 250 kilometres off the south coast of Newfoundland, Canada. Shortly afterwards a dozen transatlantic cables snapped, depriving the Canadian coast of communication.

At the time, damage was blamed directly on the earthquake. Not until 1952 did geologists Bruce Heezen and Maurice Ewing at Columbia University in New York check the times at which each cable snapped. They concluded that the earthquake had sent 200 cubic kilometres of sediment tumbling off an undersea shelf. A mixture of mud and seawater swept down at tremendous speed, snapping one cable after another. Heezen and Ewing's calculations showed that this 'turbidity current' reached speeds of up to 100 kilometres an hour, sculpting banks in the seabed and travelling 600 kilometres down the continental slope.

Today, turbidity currents are still snapping submarine cables. With 95 per cent of international phone, Internet and data transmissions buzzing through them, they can spell big trouble. With so much at stake, it's perhaps surprising that we don't have better maps of abyssal channels. Yet, with existing technologies, creating a global map would probably take hundreds of years of ship-time. For now, unknown channels are still being found when cables snap.

How underwater rivers form

Plotting channels on the map is one thing, understanding them is quite another. As you'd expect from a flow triggered by an earthquake, some are short-lived, so some underwater rivers 'dry up', just as terrestrial rivers do in a drought. They're still filled with water, of course, but there is no flow of mud and sand through them.

Some are triggered when sediment that accumulates at the top of a canyon collapses under its own weight. Others are caused by terrestrial rivers running into the sea. Take the River Congo, for example. By the time it reaches the Atlantic, its water is rich in sediment and it can trigger an underwater river that flows far into the ocean along an abyssal channel. Similarly, the Yellow River in China has created an abyssal channel where it tumbles into the Xiaolangdi Reservoir.

Although most channels seem to have been formed by terrestrial rivers, some have been spotted out in the middle of oceans, and nobody yet knows how they got there.

Muddy, rapid turbidity currents are difficult customers: instruments lowered into them are generally destroyed. The huge flow under the Bosphorus Strait is friendlier, however. It's less muddy than most. Instead of sediment, it is laden with salt from the Mediterranean. This extra-salty water is denser than the surrounding water, so it flows along the sea floor. Although the composition of the current is different from its muddy cousins, the dynamics of the flow are pretty much the same.

Jeff Peakall from the University of Leeds, UK, has studied abyssal channels around the world. He has also taken a torpedo-shaped robotic yellow submarine to the Bosphorus and made the first-ever detailed measurements of flow through an abyssal channel. He found that the underwater river twists and turns like a disturbed rattlesnake. Terrestrial rivers, of course, may wiggle or run straight depending on the surrounding terrain. But underwater rivers are different. They are strange in that all the wiggly ones are close to the equator while the straighter ones are at the poles.

Why? Peakall suspected that the culprit was the Coriolis effect, which deflects winds and ocean currents (see Chapter 6). But, unable to test different abyssal channels directly, he turned to experiments in the lab. Together with researchers at the University of Toronto, he mounted a 2-metre-long tank on a spinning table and filled it with water. On their 'seabed' the scientists built a meandering acrylic channel, and introduced dense saline flow to simulate a muddy turbidity current. Then they spun the tank at different speeds to mimic Earth's rotation at different latitudes.

They found that underwater rivers behave in dramatically different ways from their terrestrial counterparts. In both cases, the movement of water around a bend is controlled by a combination of forces. The main ones on land are the weight of the water and the centrifugal force that pushes water outwards when it flows around a bend.

But turbidity currents exist not in air, but in water. The weight of the current is countered by buoyancy, leaving the door open for the Coriolis force to have a greater influence than it does on land. This leads to some curious effects. The flow of current is deflected to one side of the channel, for example – to the right in the northern hemisphere and to the left in the southern. At the bend of a terrestrial river, surface water tends to flow outwards, while basal water flows towards the inside of the bend. But in an underwater river, this pattern is reversed.

These altered flows create unusual patterns of erosion and deposition of sediments. One bank of a channel (the side bearing the current) grows larger than the other. Near the poles, where the Coriolis force is greater, channels grow straighter than those near the equator. And while erosion pushes the bends of terrestrial rivers steadily downstream, submarine meanders stay put. Instead, once they reach a certain degree of 'wiggliness', they build up vertical levees that can reach hundreds of metres high.

While cable-laying and oil companies are interested in the strange behaviours of abyssal rivers, climate scientists are also watching closely. Undersea channels, especially those extending from terrestrial rivers, carry vast amounts of carbon in organic material and particles, much of which is eventually buried in sedimentary deposits, only re-entering the atmosphere perhaps millions of years later. Scientists are trying to quantify how much carbon is transported and buried at what rate, and how it affects the global carbon cycle. A better understanding of abyssal channels will help to model climate more accurately.

8
Life

The relationship between Earth and its living residents is intriguing. Ever since life gained a foothold about 4 billion years ago it has been 'redecorating its home'. It has profoundly changed the planet's atmosphere and the nature of its rocks. Earth's relationship with life has also spawned one of the twentieth century's most original theories.

Earth, meet life. Life, this is Earth

As things stand, if it's life you're after, there's only one planet worth visiting – Earth. Among the puzzles this uniqueness presents, none is more intriguing than how life got started. If it began here – that is, if it didn't hitch a lift from somewhere else in the universe – it means that at some point a bevy of prebiotic chemicals came together to form a thing capable of reproducing and evolving. How did that happen, and where on Earth did it all begin?

The first hints came from experiments conducted in the 1950s by Stanley Miller, a 23-year-old doctoral student then at the University of Chicago. He announced that he had made amino acids, the building blocks of proteins, with little more than a spark of electricity shot through hot air circulating in glass tubing. Miller's spark was a stand-in for primeval lightning, and the hot air, containing ammonia, hydrogen, water vapour and methane, was meant to mimic Earth's atmosphere 4 billion years ago.

Besides creating amino acids, other researchers quickly demonstrated that Miller's experiment could also produce adenine and guanine, two of the nucleic acid bases that make up RNA and DNA. But it took another 43 years for Miller to show how his experiment could produce RNA's other two bases – uracil and cytosine. He and his student Michael Robertson discovered a way for a primordial soup to make them by the bucket-load. The secret ingredient was urea, which, although produced in the original experiment, never reached high enough concentrations. Miller argued that on the early Earth urea would have reached the right levels as shallow pools evaporated – the 'drying lagoon hypothesis'.

Where life started

So, the feedstocks for carbon-based life could have been supplied by Earth's atmosphere and weather. How these were then assembled to create a viable organism is a hot area of research and in recent decades a small army of theorists and experimentalists have explored every possible chemical step. This work has uncovered a few other Earthly niches, besides drying lagoons, which could have been the cradle of life.

The earliest organism that researchers have envisaged is not what we're used to today, but something simpler. Here's why. At some point long after the development of first life, a single-celled character emerged called LUCA – the last universal common ancestor of all living things on Earth. We can tell a lot about LUCA from the common features we see in all organisms today. We know it used DNA to store recipes for proteins – genes – for

FIGURE 8.1 Could freezing and thawing of ice have played a role in creating the first life?

instance, and that those recipes were conveyed within its cell by RNA. We even know what many of those recipes were, because many vital proteins found in all cells today must have come from LUCA. And from the nature of these proteins, it is clear that LUCA used an energy-rich molecule called ATP to fuel its cellular processes, just as our cells do.

Jumping from prebiotic chemicals straight to LUCA presents a massive challenge, however. Proteins play an essential role in DNA replication, so to create LUCA from scratch would mean both proteins and DNA arising together – a stupendously unlikely coincidence. Most biologists have settled instead on an intermediate step known as the 'RNA world' hypothesis. RNA can make up genes – as DNA does – and it can catalyse chemical reactions – as proteins do. The story goes that RNA or something very like it could have enabled replication and metabolic tasks, such as energy production. Only later did DNA become the repository of genes and proteins take over metabolism.

Researchers studying the origin of life often focus on organisms that could function just with RNA. Making small RNA-like molecules from the soupy atmosphere is one thing, but how did such subunits assemble into longer strands of RNA that would have been useful as genes and catalysts? That question led Jack Szostak, now at Harvard Medical School in Boston, and colleagues to explore a soft clay mineral called montmorillonite.

The electrical character of this clay makes it attractive to short strands of RNA, encouraging them to assemble into longer chains. Szostak and his team also found that the clay can catalyse simple fatty acid precursors to form vesicles that resemble primitive cells surrounded by a membrane. A membrane is another building block essential to a living cell: it protects the contents and concentrates chemicals so that reactions take place. The team finally found that RNA strands attached to montmorillonite could be engulfed within a vesicle as it formed, creating essentially a 'protocell'.

Hence clay at the bottom of a pond might have been instrumental in kick-starting life. But it is not the only Earthly attribute to be suggested for this role. In 2013 Philipp Holliger of the MRC Laboratory of Molecular Biology in Cambridge, UK, proposed that ice may have played a role. He and his group created for the first time a long RNA catalyst that can build other RNA molecules that are longer than itself. What was surprising was that this process worked at temperatures as low as −19 °C in tiny pockets between ice crystals. In 2017 he showed how such RNA catalysts could be assembled from short strands of RNA in water subjected to freezing and thawing.

At the other extreme, Nick Lane of University College London thinks the RNA world may have emerged around undersea volcanic vents, where alkaline fluids at temperatures of up to 90 °C well up through cracks in the sea floor. Here, as the fluid hits cold seawater, minerals precipitate out of solution, forming rocky chimneys up to 60 metres high, which are replete with labyrinthine channels and pores.

Even on early Earth, these chimneys would have been rich in iron and sulphide, which can catalyse complex organic reactions. What's more, temperature gradients within the pores should have created high concentrations of organic compounds and favoured the formation of large molecules, including fat molecules and possibly RNA. Here, according to Lane, is the perfect place for the formation of self-replicating sets of RNA and membrane-like fatty vesicles.

Most intriguing of all, these vents could have generated energy via a natural proton gradient at the interface between the proton-poor alkaline vent fluid and the proton-rich seawater. This is the same kind of electrochemical gradient that drives ATP production in cells to this day, and Lane posits that the cellular machinery for generating ATP first formed in these rocky pores.

These ideas highlight the rich and complicated interactions possible between would-be life and different attrubutes of the planet. Which of them is correct is still moot and we may never uncover the answer, although plenty of research is still ongoing. There is, of course, another possibility which would make all these ideas superfluous: perhaps life never began on Earth at all...

Cosmic children

The idea that life came from space – 'panspermia' – emerged in the nineteenth century. It grew in popularity in the early 1970s, when astronomers discovered that space was replete with complex organic molecules. Panspermia seemed to answer the question of why life on Earth seemed to arise almost as soon as the planet became habitable. Could the transition from prebiotic chemistry to biology really have happened so fast?

The idea remains an unproven hypothesis on the fringes of mainstream science. Nevertheless, it still has its cheerleaders. Chandra Wickramasinghe of the University of Buckingham, UK, one of the scientists who revived panspermia in the 1970s, points to evidence such as the discovery in 2013 of microorganisms 27 kilometres up in the stratosphere. The microbes were collected during the Perseid meteor shower and were too high to have been lofted from Earth's surface.

According to Wickramasinghe, the galaxy is teeming with life, and our biosphere is just part of a vast, interconnected cosmic ecosystem. Genetic material and even living organisms are constantly exchanged between Earth and neighbouring star systems. Evolutionary change is largely driven by novel genetic material arriving from space, most likely in the form of viruses.

Others are a little less ambitious, arguing merely that life first evolved on Mars and arrived here on meteorites. Ultimately, if panspermia is ever shown to be correct, we would have to think of lifeforms on Earth not as merely terrestrial, but as children of the cosmos.

How to poison a planet

If the early Earth contributed to the formation of life, life has been repaying the debt ever since. It has changed Earth's atmosphere, oceans and geology. Nothing demonstrates life's impact more dramatically than 'the great oxygenation event' of 2.4 billon years ago, triggered by the evolution of bacteria capable of splitting water during photosynthesis.

Photosynthesis is a process for turning energy from sunlight into chemical energy for fuel and growth. Simply put, sunlight shines on a source and frees up electrons which the organism adds to carbon dioxide to create sugars and other carbohydrates. The earliest fossil evidence we have of photosynthetic microbes dates from about 3.4 billion years ago.

But photosynthesis in those early days was not of the form we think of today. Microbes would have relied on substances such as hydrogen sulfide as the source of electrons, and produced sulfur as waste.

Then, around 2.8 billion years ago, a new group of microbes emerged which employed the form of photosynthesis we are more familiar with today. This used water as the electron source and generated a highly toxic waste product – oxygen.

The gas had been all but unknown on Earth up to that point, and unwanted. It literally burned cells from the inside, grabbing electrons to create reactive chemicals called free radicals which wreak havoc within cells.

Oxygen levels surged about 2.4 billion years ago, triggering global mass murder. Most anaerobic microbes died on the spot. A few survivors found refuge deep underground or in other anoxic nooks. Today some still live deep down in swamps or inside the crypts in your guts.

In the seas, oxygen combined with dissolved iron, forming insoluble iron oxides which sank to the sea floor. We see the consequences today in vast iron-banded rock formations in Western Australia and Minnesota in the USA. As oxygen levels increased, high in the stratosphere, ultraviolet radiation from the Sun split apart oxygen molecules to create the ozone layer (see Chapter 6). This action protected life from harmful UV rays, providing a shield that would later enable life to move on to land.

The new group of photosynthesizers, probably ancestors of today's cyanobacteria, prospered in large part because using oxygen to burn carbohydrates for energy has an efficiency some 18 times that of doing it without oxygen. Life was becoming high-powered, setting the scene for the evolution of more complex life forms.

Eukaryotes capable of exploiting oxygen, arrived sometime between 2.1 and 1.6 million years ago. Their growth may have been the cause of a dip in atmospheric oxygen, but equilibrium soon re-established itself (see Figure 8.2). Later came multicellular life forms – including plants, which 'borrowed' their photosynthetic apparatus from cyanobacteria. Today, directly or indirectly, photosynthesis produces virtually all the energy used by life on Earth.

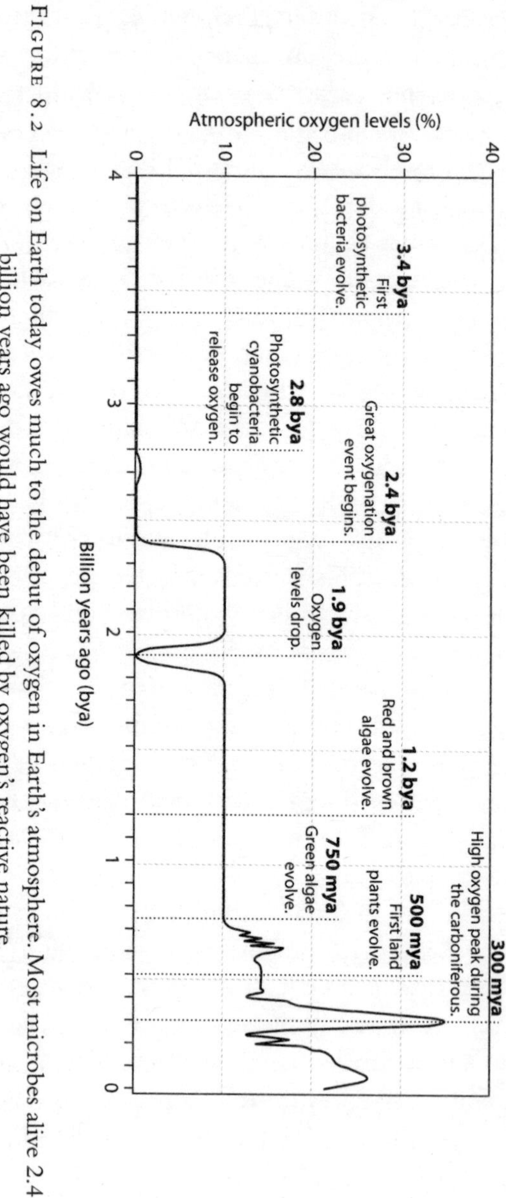

FIGURE 8.2 Life on Earth today owes much to the debut of oxygen in Earth's atmosphere. Most microbes alive 2.4 billion years ago would have been killed by oxygen's reactive nature.

The rise of Gaia

Life's interaction with the physical systems of Earth led to one of the most influential, ground-breaking scientific ideas of the twentieth century – the Gaia hypothesis, named after the ancient Greek goddess of Earth, a nurturing 'mother' of life. It is the brainchild of independent-minded scientist and inventor Jim Lovelock and is beautifully simple and appealing – proposing that the biosphere regulates itself, nurturing life in the process. It is, though, controversial with both strong supporters and strong opponents, including some who charge that it is not properly science at all, but belongs more to philosophy or even religion.

The Gaia story starts in the 1960s when Lovelock, then at NASA, and his colleague Dian Hitchcock showed that the Martian atmosphere was in a state of chemical equilibrium – a stagnant pool of carbon dioxide with a dash of nitrogen and very little oxygen, methane or hydrogen. They contrasted it with our own, which they recognized as being in chemical disequilibrium, with CO_2 and oxygen levels in constant flux. The key driver of this flux is life: photosynthesis exchanges CO_2 for oxygen, and aerobic metabolism does the opposite. Without life, our atmosphere would radically change from the oxygen-rich and life-sustaining gaseous mix we breathe to one in chemical equilibrium – one that, like the Martian atmosphere, would be inimical to life.

Earth's atmosphere is not only in flux, but is also welcoming to life, and has been for billions of years. Similarly, Earth's surface temperature, acidity and ocean chemistry seem to have been stable for billions of years, hovering around mean values that allow continued habitability. Pondering these implications, Lovelock began piecing together a novel view of life and its interaction with the planet that hosts it.

FIGURE 8.3 Jim Lovelock came up with his Gaia hypothesis after studying the 'stagnant' atmosphere on Mars.

Stated briefly, the Gaia hypothesis is that life as a whole interacts with the physical environment in such a way that it not only keeps Earth habitable but continually improves the conditions for life. It does this through a series of feedback systems similar to biological homeostasis, the mechanism by which living organisms maintain a stable internal environment. Those aspects that most affect the habitability of the planet – temperature, the chemical composition of the oceans and fresh water, and the make-up of the atmosphere – are not just influenced by life, they are controlled by it.

Within a decade of his first writings, Lovelock elevated his hypothesis to the scientifically stronger Gaia theory. In the mid-1970s he described his view as follows: 'The Gaia theory says that the temperature, oxidation state, acidity and certain aspects of the rocks and waters are kept constant, and that this homeostasis is maintained by active feedback processes operated automatically and unconsciously by the biota.'

Lovelock eventually began to refer to the planet itself as some kind of superorganism. 'The entire range of living matter from whales to viruses and from oaks to algae could be regarded as constituting a single living entity capable of maintaining Earth's atmosphere to suit its overall needs and endowed with faculties and powers far beyond those of its constituent parts,' he wrote in his 1979 book *Gaia: A New Look at Life on Earth*. In other words, Earth is not simply a planet that harbours life; it is itself alive.

The idea was elegant and quickly attracted many adherents, both scientists and non-scientists. Some researchers saw in Gaia a new way of thinking about the cycles of organic components and elements. Some followed Lovelock's lead in searching for scientific support for the idea that life regulates conditions on the planet. Some, mainly non-scientists, saw in it a new view of how humans should relate to the planet and the rest of life. Some even found the face of God in the concept.

The many faces of Gaia

There are at least three different variants of the Gaia hypothesis:

- **Optimizing Gaia**

This early interpretation remains one of the 'strongest' versions of Gaia theory. It implies that life actively controls environmental conditions, including purely physical aspects of the biosphere such as temperature, oceanic acidity and atmospheric gas composition, such that Earth remains optimally habitable.

- **Self-regulating (or homeostatic) Gaia**

This is a more recent and slightly weaker incarnation of the theory. Rather than life actively optimizing conditions on the planet, it creates negative feedback systems that keep life-constraining factors such as temperature and atmospheric oxygen and carbon dioxide levels, within certain ranges.

- **Superorganism Gaia**

Earth isn't just a physical planet that supports life; it is itself alive. This is the strongest interpretation of the theory and tends to be viewed as unscientific.

Is Gaia the right Goddess?

Gaia continues to generate scientific interest and debate: there have been several international conferences devoted to the hypothesis. The ground is shifting, though. A number of recent discoveries have cast serious doubt on the Gaia hypothesis. Two lines of research are especially damning: one comes from deep time – the study of ancient rocks – and the other from models of the future. Both overturn key Gaian predictions. In fact, if we were to choose a mythical mother figure to characterize the biosphere, it would more accurately be Medea, the murderous wife of Jason of the Argonauts. She was a sorceress, a princess – and killer of her own children.

Starting with the deep-time discoveries, one of the most powerful arguments made by Gaia proponents is that planetary temperatures remain steady and equable thanks to feedbacks that are caused, or at least abetted, by life. The single most important of these 'thermostats' is the weathering cycle (see Chapter 6). Volcanoes belch an unceasing amount of the potent greenhouse gas CO_2 into the atmosphere. Without some way of scrubbing it out, it would build up to the point where Earth would experience runaway warming that would ultimately cause the oceans to boil away – the fate of Venus some 4 billion years ago.

That scrubbing is provided largely by chemical weathering of silicate-rich rocks such as granite. This drives a chemical reaction with CO_2 that removes the gas from the atmosphere and locks it away as limestone. The rate of this reaction is increased by land plants, whose roots break up rock and allow water and CO_2 to penetrate. Plants also directly remove CO_2 from the atmosphere through photosynthesis.

So far, so Gaian. But as scientists have made ever more precise estimates of past global temperatures, the constancy predicted by Gaia theory has been found wanting. In fact, there has been a rollercoaster of temperatures, caused by the evolution of new kinds of life.

Around 2.3 billion years ago, for example, Earth endured a gigantic episode of glaciation that lasted 100 million years. It was so intense that the oceans froze completely, creating a 'snowball Earth' (see Chapter 2). The cause was life itself. The new breed of water-splitting photosynthetic microbes sucked so much heat-trapping CO_2 out of the atmosphere that the planet was plunged into the freezer.

A second episode of snowball Earth, brought about by the evolution of the first multicellular plants, happened 700 million years ago. Later still, the evolution of land plants gave the climate a double whammy. As well as reducing CO_2 by photosynthesis, their deep roots dramatically increased weathering rates. The result was that soon after the appearance of forests near the end of the Devonian period 360 million years ago, Earth entered an ice age that lasted 50 million years. The warm, verdant planet cooled rapidly and vast swathes of life died out. Not a very Gaian result.

In fact, for as long as life has existed it has been well able to devastate itself. Charles Darwin envisaged newly evolved life forms entering the world like a wedge, easing into a narrow vacant niche then expanding it gradually. Some do. But others enter like a sledgehammer, smashing away entire branches of the tree of life as they arrive.

Perhaps the worst Medean event of all was precipitated by the same biological innovation that led to the first snowball Earth: the great oxygenation event, mentioned earlier in this chapter. This unleashed a weapon of mass destruction – oxygen. Life was devastated. All that survived were the new photosynthesizers and microbes that evolved rapidly to tolerate oxygen or found anoxic refuges.

Even more damning to the Gaia hypothesis are results from the study of the mass extinctions that have occurred since the evolution of animals 565 million years ago, of which there have been five big ones and about ten more minor ones. Most of these events are now seen as 'microbial' mass extinctions, caused by huge blooms of bacteria belching poisonous hydrogen sulfide gas. These blooms thrive in the stagnant oceans that arise during intense episodes of global warming, such as the one at the end of the Permian, when prolonged volcanic activity vented vast amounts of CO_2 into the atmosphere.

According to Gaia theory, life should have buffered these events. But it did not. Far from being Gaian, their existence seems to strongly support the Medean view, as do many other events in the history of life including, arguably, the human-induced mass extinction that is going on around us now.

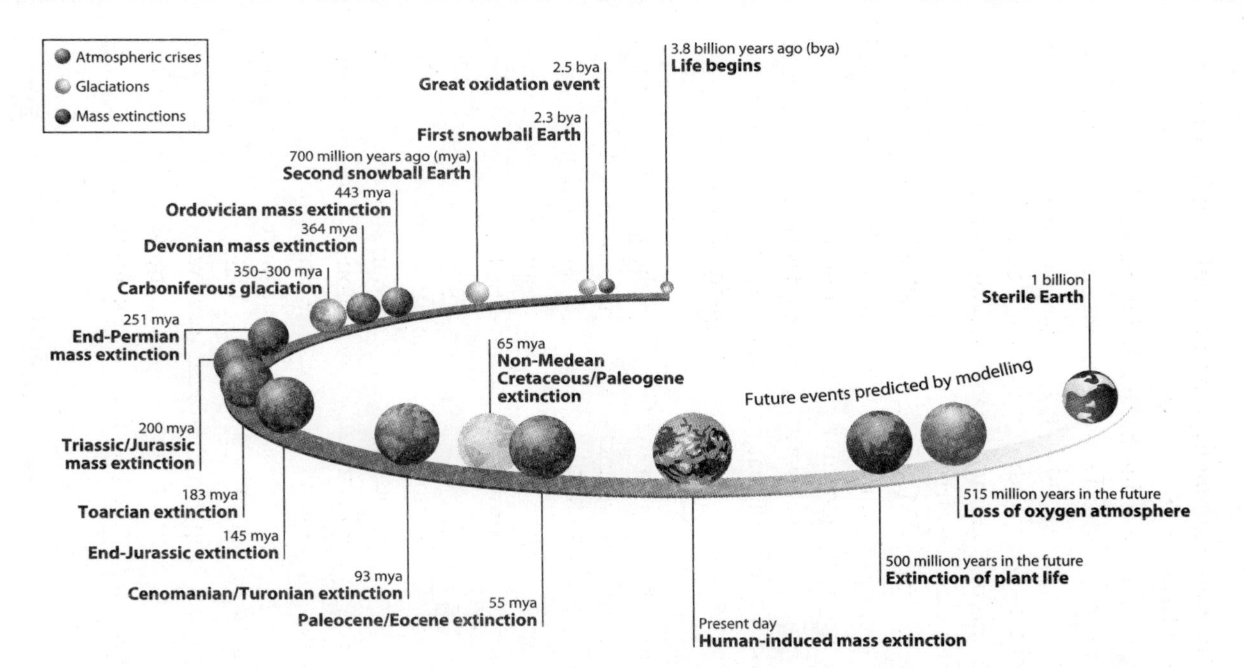

FIGURE 8.4 Almost all mass extinctions have been caused by one type of life killing another, with the notable exception of the Cretaceous–Palaeogene extinction, which is thought to have resulted from an asteroid impact.

Looking to the future

What of the future? Here, too, we can refute Gaia, and this is perhaps the most interesting – and shocking – of discoveries. Life seems to be actively pursuing its own demise, moving Earth ever closer to the inevitable day when it returns to its original state: sterile.

How so? The starting point is that the Sun is getting hotter. It has increased in brightness by about 30 per cent over the past 4.5 billion years and will carry on doing so. As the Sun burns brighter, it will cause global warming, which will increase the rate of chemical weathering. This will remove CO_2 ever faster from the atmosphere, aided and abetted by photosynthesis and plant roots.

At first, this will buffer the solar-induced temperature increase. But there will come a time – possibly as early as 500 million years from now – when there is not enough CO_2 in the atmosphere to support photosynthesis. When that calamitous day arrives, a very pronounced end of the world as we know it will begin.

The changes will be dramatic and catastrophic to life. Plants will wither and die, shutting off the main source of biological productivity and atmospheric oxygen. Animals will quickly follow. The loss of plants will also lead to a renewed build-up of CO_2 in the atmosphere, feeding a runaway greenhouse effect.

Eventually, the temperature of Earth's surface will exceed that of boiling water, and the last microbe will perish. Earth will be lifeless once more. This is very anti–Gaian, since the theory states that the presence of life on a planet should extend its habitability. The opposite is true.

If these models are correct, life on Earth is already in its old age. The adventure that started 3.8 billion years ago, and is still the only life we know of in the universe, has maybe another billion years to run. The long-term, and terminal, decline of CO_2 in the atmosphere has already started – the effect of burning fossil fuels is just a blip. Gaia is dying. Long live Medea. For now.

Putting Gaia to the test

The first review of the evidence for the Gaia hypothesis appeared in 2013 in *On Gaia: A Critical Investigation of the Relationship between Life and Earth*, a book by Toby Tyrrell, Professor of Earth System Science at the University of Southampton. One of Gaia's main arguments is that life has greatly altered Earth's environment, including the chemical composition of the atmosphere and the sea. Tyrrell found plenty of evidence of biological alteration of the global environment. For instance, life affects the planet's albedo – the degree to which Earth reflects solar energy back out to space. One way it does so is by the generation by ocean microbes of dimethyl sulfide, a chemical that rises in the atmosphere and influences the formation of clouds.

However, this effect, long held up as confirming Gaia, turns out to be a relatively weak one. And there's another catch. Life's ability to change Earth's environment could equally well support a competing idea. The 'coevolution of life and planet' hypothesis posits that life and the environment influence each other but with no requirement that the outcome improves or maintains Earth's habitability. Tyrrell found no compelling reason to favour Gaia over this alternative.

What of a mechanism for Gaia? The hypothesis would be instantly more plausible if it could be seen to arise naturally out of evolution. Tyrell found no convincing reasons to believe this, but came across fascinating cases of 'Gaias in miniature'. For instance, the interiors of termite mounds and wasps' nests are strongly thermo-regulated, experiencing much smaller day–night temperature fluctuations than the air outside. These stable internal temperatures come about partly because of how these social insects orientate their nests, but also through corrective group behaviours when brood temperatures drop too low or rise too high.

These are great examples of Gaia in action, but they do not lead us to expect that something similar must happen at the global scale. It turns out that communal regulation of a shared environment has so far been observed only in closely related individuals, whereas the global biota is the opposite – genetically extremely diverse.

Tyrrell concludes that the Gaia hypothesis is not an accurate picture of how our world works. The planet is less robustly stabilized than Gaia implies, and so more fragile. While it may be a shame that this beautiful idea doesn't hold true, he argues that it is far better that we tackle environmental issues based on an accurate view of how our Earth systems operate rather than a flawed one.

9
Welcome to the Anthropocene

Dateline, January 1950. Isaac Asimov publishes Pebble in the Sky, *his first science-fiction novel. Stevie Wonder is born. And Earth enters a new epoch – according to some scientists. The Anthropocene is named for humans and began when our numbers and activities reached such a level that we started to have a profound impact on Earth's geology, atmosphere, oceans and ecosystems.*

The epoch of humans

Since the last ice age, 11,700 years ago, Earth has been in the Holocene, the latest epoch on the geological timescale. But has that now been superseded – are we humans reshaping our planet to such an extent that we deserve our own epoch?

In September 2017 the notion of the Anthropocene took an important step closer to general acceptance. After mulling over the subject for eight years, a working group of scientists recommended adding the Anthropocene to our geological timescale.

The ultimate decision rests with the International Commission on Stratigraphy (ICS). To convince it to accept the recommendation, lots more work must be done. Somewhere near the top of the to-do list is one burning question: where in the world gives us the best view of the dawn of the Anthropocene?

For every boundary between geological periods, epochs or ages there is – or eventually will be – a Global Boundary Stratotype Section and Point, sometimes called a golden spike. It's the place that researchers have settled on as providing the best snapshot of our world lurching from one named chunk of geological time into the next (see Chapter 2).

If you want the best view of the dinosaur-filled Cretaceous giving way to the Palaeocene, for instance, the official advice is to go to El Kef in Tunisia. For the moment when the largest mass extinction in history eased us from the Permian period into the Triassic, head to Zhejiang Province in China.

So where might geologists one day single out as best marking the beginning of the Anthropocene? The net is initially being cast widely, according to Jan Zalasiewicz at the University of Leicester, UK, who is convenor of the ICS's Anthropocene Working Group. There are lots of potential environments out there, including lakes, anoxic marine basins and ice strata.

The fact that the suggested shift into the Anthropocene is so recent might make the decision harder. Ideally, the golden spike will be a site where something – like ice or sediment – is accumulating steadily enough to give us a recognizable year-by-year record. It also must be a place where this record isn't in danger of being destroyed by either natural erosion or human activity.

There's another consideration: what kind of 'marker' best demonstrates the Anthropocene's arrival? Geological boundaries are usually tied to a global event rather than a raw date. The end of the Cretaceous, for example, is defined by a widespread extraterrestrial iridium layer and a series of extinctions generated by an asteroid impact. All this is can be clearly seen in the sediment layers at El Kef.

FIGURE 9.1 Atomic bomb tests appear to have coincided with the start of the Anthropocene.

Capturing the Anthropocene's golden spike

The dawn of the Anthropocene could be tied to a spike in radioactive plutonium from the fallout of the numerous atomic bomb tests in the mid-twentieth century. Or perhaps it could be marked by the first appearance of plastic in the layers of mud accumulating on ocean floors and lake beds.

It's possible that many signals will have their role to play. For example, the forerunner of the proposed Anthropocene, the Holocene, has as its golden spike a core from the Greenland ice sheet which preserves an isotopic signature of global warming at the end of the last ice age. Auxiliary sites around the world also exist, including a sediment core from a lake in Japan which captured pollen changes associated with the rise in global temperature.

But that still leaves us with our initial question. Selecting sites that highlight different aspects of the Anthropocene doesn't give us a definitive golden spike. So where in the world would it be?

Colin Waters at the British Geological Survey, another member of the Anthropocene Working Group, thinks that the Santa Barbara Basin off the coast of California, and the Cariaco Basin off Venezuela, are worth considering. Both accumulate sediment at a steady rate and may well contain a record of plutonium. They're also unlikely to be disturbed by human activity or other processes.

Another option might be to go with a limestone speleothem, such as a stalagmite or stalactite, in a cave. This would offer a record of carbon isotopes that captures the atmospheric impact of human activity. It's also possible that the Anthropocene, like the Holocene, will best be captured in an ice core from Greenland.

Perhaps the most poignant place on Earth to mark the dawn of the Anthropocene would be on a coral reef. Zalasiewicz reckons that the annual growth rings in a particularly fast-growing coral like *Porites* should provide a high-resolution – and rock-solid – chemical record of the moment the Holocene gave rise to the Anthropocene.

We are all too familiar with the threats that coral reefs face as a consequence of global warming and ocean acidification. Waters reckons there would be something poetic in choosing a coral – plucked from somewhere like the Caribbean to be stored in a museum – as the one object on Earth that best records the moment intense human activity began to change the whole planet.

Markers of human influence

Even if humanity is long gone in tens of millions of years, there will still be clear signs of our existence preserved in our planet's geological record. As we've heard, the deposition of plutonium and other radioactive isotopes coincides with the start of the period when humans began to have a large, widespread impact on the planet. And there are other potential markers...

Fossil fuels

The products of burning fossil fuels will be an obvious giveaway of the Anthropocene. Current rates of carbon emission are thought to be higher than at any time in the last 65 million years. The concentration of carbon dioxide in the atmosphere has risen sharply since 1850 and is now around 410 parts per million, which will be recorded in cores taken from any ice sheets in Greenland or Antarctica that survive global warming. Burning fossil fuels has also increased the isotopic ratio of carbon-12 to carbon-13, which will be detectable in tree rings, limestone, and fossilized bones and shells. Our fuel consumption also spreads small particles of carbon into the air, which can become captured in sediments and glacial ice.

New materials

One of the biggest signs of our time will be the presence of three things we use every day: concrete, plastics and aluminium. Aluminium in its elemental form was unknown before the nineteenth century, but we have now produced around 500 million tonnes of it. Concrete has been around for longer – it was invented by the Romans – but in the twentieth century it became our most widely used building material. We have now produced about 50 billion tonnes of the stuff – enough to spread a kilogram on every square metre of Earth – and more than half of that was made since the 1990s. Production of plastics has grown rapidly since the 1950s, and we now produce more than 300 million tonnes a year. Sediments containing any of these materials will be a clear sign of the Anthropocene.

Changed geology

Every time we destroy a patch of rainforest, it changes the future of Earth's geology. So far, we have transformed more than 50 per cent of Earth's land area for our own purposes. Deforestation, farming, drilling, mining, landfills, dam-building and coastal reclamation are disturbing sedimentary processes on a wide scale, disrupting how layers of rock are laid down.

Fertilizers

Our attempts to feed a burgeoning population also leaves clear markers. Levels of nitrogen and phosphorus in soils have doubled in the last century because of our increased use of fertilizers. We produce 23.5 million tonnes of phosphorus a year, twice the rate seen during the Holocene. Human activity has had perhaps the biggest impact on the nitrogen cycle for 2.5 billion years, increasing the amount of reactive nitrogen – that's everything except unreactive nitrogen gas (N_2) – by 120 per cent compared to the Holocene.

Global warming

Anthropogenic climate change will be easily distinguishable in the future. In the twentieth century, Earth's temperature rose by between 0.6 and 0.9 °C, more than the amount of natural variability seen in the Holocene, which has been calculated based on the oxygen isotopes in Greenland's ice cores. Average global sea levels are higher than at any point in the past 115,000 years and are rising rapidly – a fact that may also be detectable in the future

Mass extinction

For as long as life has existed, organisms have gone extinct, but mass extinctions sparked by large-scale global changes mark the end and beginning of several geological periods. Evidence is growing that widespread human activity has triggered the sixth big mass extinction event in Earth's history, with three-quarters of species set to be wiped out in the coming centuries. Palaeontologists of the future will notice the sudden disappearance of many species from the fossil record as the Anthropocene gets under way.

Plastic pestilence

As well as leaving their fingerprints in sediments for future geologists to find, several markers of the Anthropocene represent real threats to Earth's systems today. One of these has emerged only since the late 1990s. Plastics are undoubtedly very useful materials, yet our apparent addiction to them, combined with a reluctance to recycle has created a problem of global proportions. It's led some to jokingly suggest that the new human epoch should really be called the 'Plasticene'.

Of the 320 million tonnes of plastics produced annually, about a third is thrown away soon after use. Much is buried in landfill, where it will probably remain, but a huge amount ends up in the oceans. Some of this is washed up on beaches or eaten by wildlife. Most remains in the sea, where it breaks down into ever smaller fragments. However, our knowledge of its ultimate fate is hazy. We are only just beginning to understand how much plastic pollution is choking the seas and its impact on the health of sea creatures and those who eat them. We have still to discover where the stuff will end up in the distant future – whether plastic debris will break down entirely or leave a permanent scar.

The scale of our plastic problem became clear in 1997, when US oceanographer Charles Moore came across a huge area of floating trash – now dubbed the 'Great Pacific Garbage Patch' – as he sailed from Hawaii to California. It was soon found that other oceans contained similar concentrations of rubbish.

These patches are created by surface currents, or gyres, which turn in great circles on either side of the equator (see Chapter 7). And just as noodles gather in the centre of a bowl of stirred soup, anything caught in these currents is likely to drift into the middle. The five biggest concentrations of marine debris are in the Indian Ocean, the North and South Pacific and North and South Atlantic (see Figure 9.2). In 2014 Moore reported finding one spot in the Pacific gyre where there was so much rubbish he could walk on it.

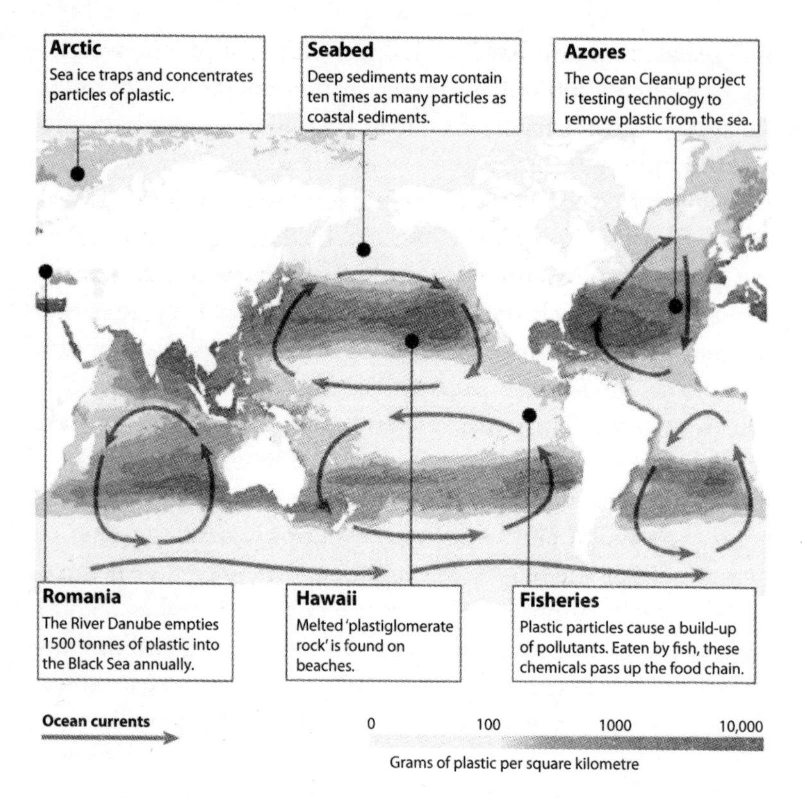

FIGURE 9.2 Much of the oceans' plastic waste is found near heavily populated coastlines, but farther out it is concentrated in five 'gyres'.

Getting a measure of how much plastic is polluting the seas is difficult. An international team headed by Marcus Eriksen at the Five Gyres Institute in Santa Monica, California, gathered data on the amount of plastic caught in nets towed behind research ships over a period of six years. This was added to records from spotters who stood on the decks of these ships and counted every piece of plastic they observed.

The team estimates that 5.25 trillion pieces of plastic, weighing a total of 260,000 tonnes, are floating at sea. Most is big stuff like buckets, bottles, bags, disposable packaging and polystyrene foam.

According to the industry trade body Plastics Europe, global production hit 322 million tonnes in 2015. Given that it's often cheaper for manufacturers to produce virgin material than to recycle plastic, much of this material is thrown away after use. So our best figures for floating plastic amount to less than 0.1 per cent of the plastic produced each year.

This raises an intriguing question – why haven't we found more? Where is all the plastic going? The answer could be that plastic breaks down more quickly than we thought, as the action of sunlight and waves degrades it into small fragments. The missing plastic may exist as a soup of tiny pieces suspended in the water column. Eriksen's team reckons there are 35,500 tonnes of plastic particles measuring less than 5 millimetres across. But this figure seems low.

There are a few possible explanations. Plastic particles less than a third of a millimetre across will slip through the trawl nets because the mesh size is too large, so a huge amount of plastic could have been overlooked.

Marine biologist Richard Thompson of Plymouth University, UK, thinks plenty of plastic may be locked up in ice. In June 2014 his team reported finding up to 234 particles of plastic per cubic metre of Arctic sea ice – several orders of magnitude higher than in the heavily contaminated waters of the gyres. He suggests that as seawater turns to freshwater ice, it traps and concentrates small particles. Given that there are about 6 million square kilometres of sea ice, this could represent a huge reservoir of plastic. If the ice melts, this material will be released back into the sea.

Thompson and his team have also discovered another place where plastic is accumulating. They have published data showing that tiny pieces of plastic and other polymers, mostly in the form of fibres, are up to 10,000 times more abundant in deep-sea sediments in the Atlantic Ocean, the Mediterranean Sea and the Indian Ocean than in surface waters. Samples contained as much as 800,000 particles per cubic metre. The number of samples – just 12 sediment cores and four chunks of coral – was small, but plastic debris was present in them all.

Eat your plastics

We have still to discover the full impact of tiny plastic fragments on marine creatures and the wider food chain. We know that larger animals, including birds, turtles, fish and whales, confuse plastic trash with food (see Figure 9.3). They choke to death or die of starvation as their stomachs become clogged. But the effect on smaller sea dwellers is far more complex.

For some microbes, plastic is the equivalent of a hotel buffet table. Any hard surface in the ocean tends to become a collection plate for nutrients. This is why the huge rafts of floating plastic are attracting an entire cast of characters, creating a new ecosystem dubbed 'the plastosphere'.

Among these characters are bacteria of the genus *Vibrio*, which includes several pathogenic species, not least the bug that causes cholera. Viruses might also find a home on the plastic, which would not be surprising since there are far higher concentrations of viruses in the water column than there are microbial cells.

There is evidence, too, that plastic microparticles are entering the food chain. Even without viruses and bacteria, microplastics aren't good news for fish. They can reduce the efficiency of food absorption and, as they break down, release additives such as flame retardants, which are toxic, and phthalates and bisphenol A, which can mimic hormones. Plastics also act like sponges for chemicals in seawater, absorbing organic pollutants including polychlorinated biphenyls (PCBs), and pesticides such as DDT. Studies suggest that pollutants stuck to plastics can poison fish.

For human fish eaters, the message is obvious: beware!

FIGURE 9.3 Plastic kills large marine mammals by trapping, choking or starving them. Its effects on other ocean-dwellers are more uncertain.

Stemming the flow

Huge amounts of plastic enter the oceans via rivers. Major components of this waste are fibres from synthetic clothes released during washing and microbeads used in many cosmetics. Waste treatment plants can't filter them out, so they end up in rivers.

In 2014 the state of Illinois passed the world's first ban on microbeads, after studies showed that the tiny plastic spheres are a common pollutant floating on the surface of the Great Lakes. Since then, national governments, led by Canada's, have started to introduce bans. Some manufacturers have also acted: Unilever, Colgate-Palmolive, Procter & Gamble and Johnson & Johnson have all committed to eliminating the beads from their products.

Slowly, the tide is turning against other plastic products. In several countries plastic bags now come with a price tag aimed at deterring shoppers from using them. In 2017 the UK government announced an inquiry into how best to control amounts of 'single-use' containers, such as takeaway cartons and drinks bottles. In January 2018 it banned production of microbeads in cosmetics and cleaning products in England, with a ban on sales to follow.

Meanwhile, some groups are hoping to harvest plastic from the gyres. In 2014 an organization called The Ocean Cleanup completed a trial of a floating boom system in the Atlantic near the Azores. Based on the results, the group estimates that floating debris in a single gyre could be cleared in five to ten years without harming wildlife. It plans to start a global clean-up in 2020.

10
Climate change

Rapid climate change is not only a potential marker for the start of the Anthropocene: it is making its presence felt in worrying ways today. The idea that humans are to blame is still controversial in some circles, so what does science have to say on the matter? Are we to blame, can we stop global temperatures rising further, and, if not, what are the likely consequences?

The writing on the wall

One thing is certain, the Anthropocene will be marked by rapid climate change. Earth's average surface temperature has risen more than 1 °C since the Industrial Revolution began – a pace that far exceeds any natural variability seen through the Holocene. Average global sea levels, which had been relatively stable over the past few thousand years, have started to rise at an accelerating rate.

It is impossible to predict exactly what impact climate change will have, mainly because we don't know how people will alter the environment in the future. Yet what science has revealed so far gives us a good idea of what could happen in various scenarios.

Greenhouse gases are warming the planet

From melting glaciers and earlier springs to advancing tree-lines and changing animal ranges, many lines of evidence back up what thermometers tell us – Earth is getting warmer.

There are two potential explanations for this: more heat is reaching Earth, or less is escaping. The first option can be ruled out. The amount of energy from the Sun entering Earth's atmosphere varies by about 0.1 per cent over a typical 11-year solar cycle, but satellite data show no increase in this value that can account for the rising temperatures of recent decades. We are left with the second possibility: less heat is escaping.

There are several reasons why this could be so. One is a rise in greenhouse gases such as carbon dioxide and methane. These gases absorb specific frequencies of infrared radiation – heat – that would otherwise escape into space (see Chapter 6). They re-radiate some of that energy towards Earth's surface and lower atmosphere. Higher greenhouse gas levels cool the upper atmosphere, meaning that less heat is radiated into space and the planet warms.

Since the beginning of the industrial age in the nineteenth century, CO_2 levels in the atmosphere have increased from 280 parts per million to 408 parts per million in 2017. Satellite measurements now show that less infrared of the frequencies absorbed by CO_2 and other greenhouses gases is escaping the planet, and that more of it is being reflected back to Earth's surface.

Further evidence comes from studies of Earth's past climate, which reveal that the planet has warmed whenever CO_2 levels have risen. While many factors affect our planet's climate, there is overwhelming evidence that rising CO_2 levels are the prime cause of the recent warming.

Other pollutants are cooling the planet

We pump all kinds of substances into the atmosphere. Nitrous oxide and CFCs warm the planet as CO_2 does. Black carbon – soot – warms things up overall by soaking up heat, but cools Earth's surface by shading it. Yet other pollutants reflect the sun's heat back into space and so cool things down.

After large volcanic eruptions that pump sulfur dioxide high into the atmosphere, such as that of Mount Pinatubo in the Philippines in 1991, the planet cools a little for a year or two. But unlike CO_2, the influence of SO_2 is short-lived because in air, SO_2 forms an aerosol of tiny droplets that soon rain out.

The burning of sulfurous fossil fuels has been adding huge amounts of SO_2 to the atmosphere. Between the 1940s and 1970s this pollution was so high that it partly balanced out the warming effect of CO_2. But as Western countries cut their sulfur emissions to tackle acid rain, the masking effect declined.

Sulfur emissions began rising again in 2000, largely because China built more coal-fired power stations. Now China is cleaning up its act but sulfur emissions are rising in other developing countries.

The planet is going to get a lot hotter

Doubling atmospheric CO_2 on a planet with no water or life would warm it by about 1.2 °C. Even without the complicating effects of aerosols, things aren't that simple on Earth.

Take water. Water vapour itself is a powerful greenhouse gas and when an atmosphere warms, it can hold more of the stuff. As soon as more CO_2 enters a watery planet's atmosphere, its warming effect is rapidly amplified.

This is not the only such 'positive feedback' effect. Warming leads to the rapid loss of snow cover and sea ice, both of which reflect sunlight back into space. The result is that more heat is absorbed and warming escalates. Longer timescales bring changes in vegetation that also affect heat absorption. Vast ice sheets can melt away, further decreasing the planet's reflectivity. Barring some catastrophe such as a megavolcano eruption, the planet is going to warm considerably. But by how much?

One way to get an idea of how complex feedbacks play out in Earth's climate is to use computer models. The other method is to look at how changes in CO_2 have affected our climate in the past. Both methods have limitations. It's hard to work out what the climate was like in the past, for example, and CO_2 levels have never risen as rapidly as they are rising now.

Climate models and studies of the recent past suggest that doubling CO_2 levels warms the planet by around 3 °C – a yardstick known as 'climate sensitivity'. Studies of the distant past, however, point to sensitivities of 6 °C or more. One reason for this discrepancy is that climate models include only 'fast' feedbacks such as changes in clouds. They leave out longer-term feedbacks such as changes in ice-sheet coverage.

This means that models might be giving us a good idea of how much warming there will be over the next few decades, but underestimating warming on longer timescales. Studies suggest that if business continues as usual the planet will warm by 5 °C by 2080 to 2100. With big cuts to greenhouse gas emissions, warming could be limited to about 3 °C. As of winter 2017, it appears we are heading for somewhere between these two figures.

Sea level is going to rise many metres

When oceans warm, they expand. When ice on land melts or slides into the sea, that also pushes levels up. If all the ice in

FIGURE 10.1 Water will expand and ice sheets melt in response to global warming. The next century looks to be extremely dangerous for people living in low-lying coastal areas.

Greenland and Antarctica melted, sea level would rise more than 60 metres.

Today, we are in a warm period at the end of an ice age. In comparable interglacials in the past half million years, when temperatures were less than 1 °C warmer than they are now, sea level was around 5 metres higher than it is today. Around 3 million years ago, when temperatures were just 1 to 2 °C higher than preindustrial temperatures, sea level was at least 25 metres higher than at present.

So even relatively small increases in temperature can eventually lead to huge sea-level rises. The trillion-dollar question is how long it will take. Until recently, it was thought that it would be many centuries before there was any significant melting of the great ice sheets of Greenland and Antarctica. But observations show that they are responding far faster than we expected. Ice-sheet models also forecast rapid ice loss. Based on these findings, one 2017 study concluded that sea level could rise by as much as 3 metres by 2100.

What's also alarming is that these studies are revealing that in places, once these ice sheets start melting the process is irreversible (barring another ice age). It may already be too late to stop the seas rising at least 5 metres, and we could soon be committed to an eventual rise of 20 metres.

Why weather is running wild

Sea-level rise is going to affect many millions of people who live in low-lying coastal plains and cities. For everyone else, the most immediate impact of climate change is going to be on the weather. And we do seem to be seeing a lot of extreme weather.

Etched into the memory of many Europeans is the heatwave of 2003, which killed tens of thousands. Russia suffered an even worse heatwave in 2010, blamed for an astonishing 70,000 deaths. In 2012 the US 'summer in March' saw temperatures surge in the central northern states of the USA. Pellston, Michigan saw temperatures hit 29 °C, vaporizing the previous record by more than 17 degrees. The hyperactive hurricane season of 2017 was the costliest on record.

Climate scientists have long warned that global warming will lead to more heatwaves, droughts and floods. Yet some recent extremes, such as the summer in March, are way beyond the predictions of our climate models. And there have been extremes of cold as well. In Rome, ancient monuments crumbled when a big freeze hit Europe in February 2012. On the northern edge of the Sahara Desert, the streets of Tripoli in Libya were blanketed with snow. Record cold weather struck East Asia in January 2016: in Japan, Okinawa saw sleet for the first time on record and 85 people died of hypothermia and heart attacks in Taiwan.

Studies show that thanks to human activity our weather is getting wilder – not only steadily hotter but also more variable. Is this just a blip, or are we in for ever more freakish weather as global warming increases?

Even in an unchanging climate, our weather varies a lot. Each summer will be different. Take the average summer temperature every year, and you will get a series of numbers scattered about a long-term mean, distributed in a pattern more or less like a bell curve. Wait long enough, and you will sweat though a few very hot summers and grumble through a few very cool summers.

Over the past century the surface temperature of the planet has increased by 0.8 °C on average, which has shifted the familiar range of weather into warmer territory. Cooler summers have become less likely and warmer summers more likely. Contrary to what you might think, this kind of shift increases the odds of extremely hot summers by more than it increases the odds of slightly warmer summers (see Figure 10.2).

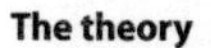

The theory

In a constant climate, temperatures should fit a bell curve – average temperatures are most likely and extremes of hot and cold are rare.

If the climate warms, this probability distribution will shift. Even in the simplest scenario, if the distribution shifts but the shape remains the same, the probability of moderate heat increases slightly while the probability of extreme heat increases greatly.

In theory, the distribution could not only shift but also widen, if weather becomes more variable as it warms. This is worse, as it means there will be an even greater increase in the probability of extreme heat, yet extreme cold will still occur occasionally, too.

What's really happening

Land temperatures over the northern hemisphere show that the bell curve is both shifting and widening as the planet warms.

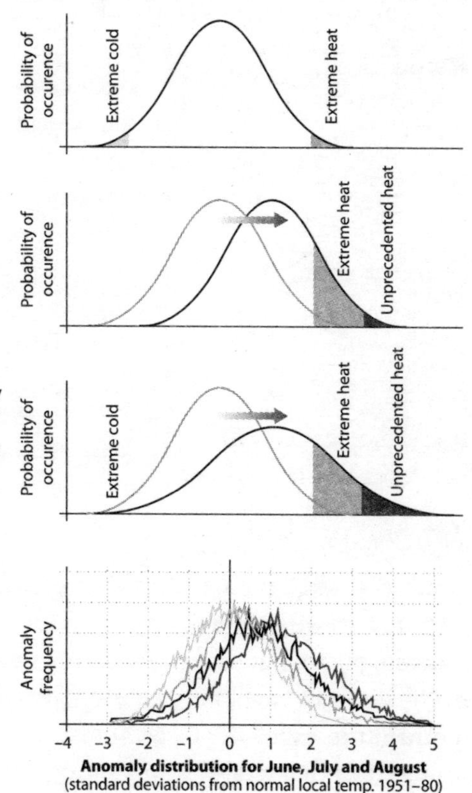

1971–81 1981–91
1991–2001 2001–11

FIGURE 10.2 The way the weather is developing looks as though we will suffer more bouts of extreme heat, but still punctuated by periods of extreme cold.

Damper downpours

Rising temperatures are leading to all manner of extreme weather. The capacity of air to hold more moisture increases exponentially as its temperatures rises. This means that when rain falls it can become a deluge, increasing the chance of catastrophic floods.

Floods are not the only result. When water vapour condenses to form clouds, it releases latent heat, and this heat is what powers most kinds of storms, from thunderstorms to hurricanes (see Chapter 6).

In 2014, the UN Intergovernmental Panel on Climate Change noted that human influence is linked to increases in the number of heavy precipitation events. Theory also suggests that any storms that do occur will tend to be more powerful because there is more heat available to them. As we saw in the Caribbean in 2017, the damage done by storms rises rapidly as wind speeds increase.

Beyond predictions

Simple physics, then, tells us that global warming should make extreme weather more extreme, from stronger storms to hotter heatwaves, drier droughts and damper downpours. This is indeed what has been happening – except that, in recent years, the magnitude of some of the record breakers has been jaw-dropping.

In 2003 temperatures in Europe were much higher than in any summer for at least 500 years. Stefan Rahmstorf of the Potsdam Institute in Germany points out that in Switzerland the average summer temperature broke the previous record by 2.4 °C. It is not unusual for the records for particular days to be broken by fairly wide margins, but for the average of an entire season to be so much warmer is extraordinary. The ferocity of Russia's 2010 heatwave was equally surprising.

There is little doubt that things are going to get even worse. What is especially worrying, though, is that the rise in extremes can't be accounted for solely by the warming we've recorded so far. Events like the 2003 and 2010 heatwaves were projected to occur only after much greater warming, towards the end of this century. And while one or two freak events might be dismissed as simple bad luck, there have been suspiciously many of them in recent years.

James Hansen, who heads the climate science, awareness and solutions programme at Columbia University's Earth Institute in New York State, analysed records of local temperatures across the globe, in each case totting up June, July and August to get an overall temperature for this period. The results show that, relative to the period 1951 to 1980, an increasing area of the planet's surface is experiencing highly anomalous heat extremes each year.

To a large extent, this is just what is expected in a warming world. However, Hansen's analysis shows that there is more to it than that. The weather is getting not only warmer, but also more variable. Between 1951 and 1980, the average range in local summer temperatures across the entire globe was 0.55 °C; from 1981 to 2010, it had gone up to 0.58 °C. Some locations, especially those far from the stabilizing influence of the ocean, see much more variability and more increase. Project that into the future, and we already have more cause for concern than we had with a mere rise in mean temperature.

Slowing the jet streams

The weird weather we are experiencing isn't just about rising temperatures. You don't need to be a climate expert to conclude that a heatwave did not cause snow in Tripoli. Yet some researchers think they know what might be to blame for that – a lazy jet stream.

Jet streams are high-speed winds that carve a snaking path through the atmosphere at altitudes of between 7 and 12 kilometres (see Chapter 6). The strongest are the two polar jet streams, one in each hemisphere, which are driven by the difference in temperature between the warm tropics and cold poles. In the tropics, the atmosphere is puffed up by higher temperatures: Jennifer Francis of Rutgers University in New Jersey says that it is as though there is a hill from the tropics tilted down towards the poles.

Gravity pulls some of this air down towards the poles. And because of Earth's spin, the air gets deflected off to one side, which is what drives the polar jet streams from west to the east.

The positions of the jet streams aren't fixed. They move around, shifting south or north and also developing big

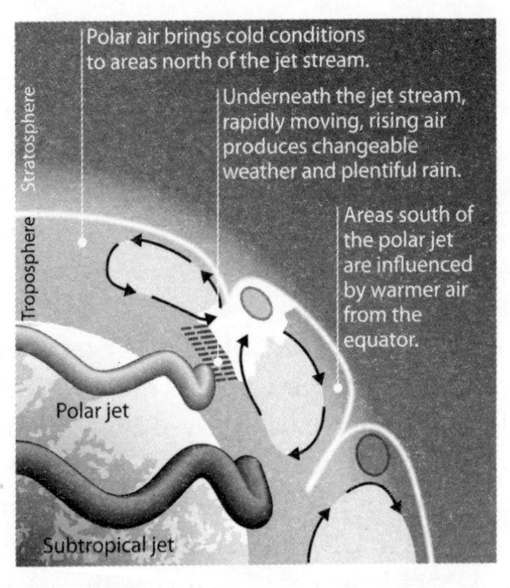

FIGURE 10.3 The position of the polar jet stream determines the weather at mid-latitudes in the northern hemisphere.

meanders, or waves. Humanity is now interfering with this vital component of the atmosphere. The Arctic is warming far faster than the rest of the planet, in part because its sunlight-reflecting snow and ice are melting to expose dark, sunlight-absorbing land and sea. This is reducing the temperature difference between the tropics and the Arctic. In 2009 Francis showed that in summers with less sea ice in the Arctic – meaning more heat being absorbed by the ocean – the atmospheric hill had a gentler slope. The upshot is that the engine driving the northern polar jet stream is weakening.

As the jet stream slows down, it takes a mazier path, with meanders that move around more slowly. This is crucial, because the jet stream pushes the weather systems that we experience around the globe. So when a stream's position changes more slowly or stays in one place for weeks – what meteorologists call a blocking pattern – the weather is more likely to become extreme.

If the jet stream shepherds one low-pressure system after another towards you, then you will soak – as happened in 2012 to the UK, which saw widespread flooding in the wettest April in 100 years. Likewise, bits of the globe can get stuck under a vast tongue either of hot, dry air stretching north from the tropics in summer, or of ice-cold air reaching south from the Arctic in winter.

A couple of days of hot or cold, wet or dry doesn't matter much to most people. But a couple of weeks can matter a great deal. Blocking patterns have played a part in much of the extreme weather around the northern hemisphere in recent years, including some of the freezing winter weather, Europe's 2003 heatwave and the US summer in March of 2012.

Other researchers have confirmed that the jet stream has been weakening, and shown that this leads to more blocking events. Now Francis has found another effect of the warming Arctic: warming the north more than the south stretches the northern peaks of high-pressure ridges farther northward. Again, that makes the meanders of the jet stream more extreme, bringing warm air further north, and cold air further south – to places such as Rome and Tripoli.

Provoking the elements

There could well be other, as yet unidentified mechanisms contributing to the wildness of our weather now, or which might kick in as the world warms further. For example, the ocean joins with the atmosphere in a roughly periodic pattern called the El Niño Southern Oscillation, or ENSO, in which warm water sloshes back and forth across the surface of the Pacific Ocean partly in response to changes in the trade winds.

One part of that oscillation, El Niño, occurs when warm Pacific water spreads eastwards. It warms the globe and has widespread knock-on effects. The El Niño that took place between 2014 and 2016, was a big one, worsening droughts in Ethiopia and Somalia and contributing to the driest growing season for 35 years in southern Africa. Vietnam, Indonesia and many islands across the Pacific also grappled with El Niño-influenced droughts. According to the UN, the event increased hunger and suffering for 60 million people.

A big question is whether things will get even worse as temperatures rises. What if ENSO and other climatic oscillations don't just continue as before but, as the world warms, become even greater in magnitude? There's no evidence one way or the other on this yet, says Stefan Rahmstorf, although he points out that, since we are changing the whole energy balance of the climate system, it would be surprising if these patterns of variation did not change.

FIGURE 10.4 Clouds are big players in keeping Earth cool. How they will react to increasing global temperatures is a critical question.

The great cloud conundrum

As we have seen in Chapter 6, clouds are enigmatic and difficult to study. These characteristics also present a challenge to climate-change scientists. At present, the overall effect of clouds is as a global heat shield, reflecting sunlight that would otherwise bake Earth and obliterate life. The big question is what will happen to this heat shield as the planet warms.

It might grow a little stronger, reflecting more light and slowing the warming. Or it could weaken, meaning that the world will warm even faster. Which way the pendulum swings is crucial because it could mean the difference between a planet that is 3 °C hotter next century – very bad but probably survivable – or 6 °C – which would be catastrophic.

All clouds trap heat beneath them in the form of long-wave infrared radiation. This is why temperatures fall less on cloudy nights than on clear ones. But clouds also reflect some sunlight straight back into space and, less obviously, act as radiators, emitting infrared to space from their tops. So a cloud is a parasol, a blanket and a cooling fin all at once (see Figure 10.5).

The overall impact of clouds depends on their height and type. Low clouds cool the planet: although they trap some heat, they also reflect a lot, and their fairly warm cloud tops emit a lot of heat to space. High clouds emit much less from their colder cloud tops, and often reflect little too, so they help to warm the planet.

Low cloud is more widespread than high, which is why clouds cool the planet overall. In fact, if you were to strip away all clouds, it might lead to a runaway greenhouse effect that would eventually boil away the oceans, according to calculations by Colin Goldblatt at the University of Victoria in British Columbia, Canada.

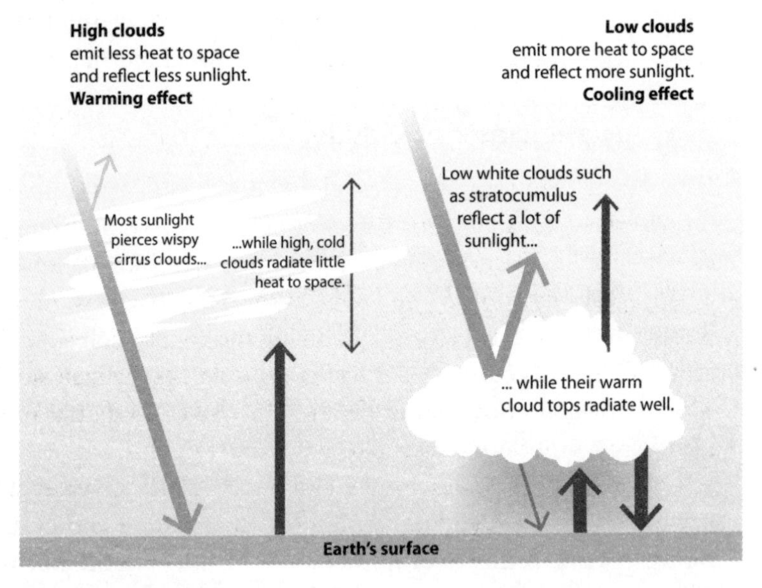

FIGURE 10.5 How high in the atmosphere clouds form determines whether they have a net cooling or warming effect on the planet.

That's not going to happen, but what will come to pass in a warmer world is proving difficult to pin down. The best way to find out, you might think, would be look at how clouds have changed over the past century as the planet warmed. This turns out to be extremely tricky.

Every approach to cloud observation has shortcomings. Weather stations on land are no use for the more widespread ocean clouds. Observations from ships are patchy and subjective. Instrument-laden planes are scarce. Weather satellites give some insights, but drift and decaying orbits plague their data. And the dedicated climate satellites of NASA's Earth Observing System have only been watching clouds for a decade or so, not long enough to catch long-term trends.

Even if we did have a good global record of cloud behaviour, it might not be a reliable guide to what happens when the planet gets warmer. As the temperature soars, we might pass some threshold that produces big changes in cloud behaviour.

The opaque nature of clouds

If we understood exactly how clouds work, we could predict future behaviour in a climate model. But cloud computing isn't easy. The inner workings of a cloud involve turbulent flows of air on scales ranging from a few kilometres to a few metres. This is invisibly small to global climate models, which slice the atmosphere into cubes 100 kilometres wide. Specialized small-scale models can now capture eddies down to 100 metres or so, but these cannot encompass large weather systems.

On even finer scales inside clouds, droplets of water and crystals of ice are colliding, coalescing, condensing and evaporating. Much of this microphysics is well understood, but not all of it. Zoom in even more, and you see that clouds cannot form without a fine mist of aerosols around which water can condense or freeze (see Chapter 6). With more particles you may get a whiter, longer-lived cloud, making a better parasol.

Models cannot capture all these processes, so they have to rely on approximations, such as the observed relationship between cloudiness and, say, humidity or temperature. These relationships can then be plugged into the models. But as we have seen, observations are not perfect, so we have no universal relationship between all the properties of the atmosphere and the amount and type of cloud we should get.

This leaves modellers in a difficult place. For example, cloud cover correlates well with the temperature difference between ground level and 3 kilometres up. But it correlates equally well with another measure that includes both temperature and humidity. Unfortunately, models give completely different predictions for what happens when Earth warms up, depending on which option is chosen.

Despite these difficulties, there has been progress with some types of cloud. Models and observations agree that high clouds will, on average, be pushed higher still as temperatures rise. That makes their cloud tops even colder, so they become less effective at radiating heat. Meanwhile, storm tracks will probably shift towards the poles, where clouds reflect less solar heat. Both of these factors will act to amplify warming.

The tropics' cooling blanket

A major part of the global heat shield is found in the tropics and subtropics, where vast expanses of low stratocumulus cloud stretch over large parts of the oceans on most days. Scientists have found these clouds exert a constant, powerful cooling influence over our climate. Yet, here again, the models clash. Some predict almost no change in these low clouds as temperature rises, others a sharp decline that amplifies global warming.

This has led to deeper and deeper exploration of the physical mechanisms taking place above those tropical waters. Some look like good news: these are 'negative feedbacks' which act to slow warming as temperature rises. For example, where warm dry air descends towards tropical oceans as part of global circulation patterns, it can trap sheets of low, cooling stratocumulus cloud. With warm air above and cooler air below – a temperature inversion – the clouds cannot rise and lose their moisture by raining. And as global temperatures rise, the warm downdrafts should get warmer, strengthening the inversion effect and increasing cloud cover on average.

At least, that is what observations and small-scale models suggest, according to work by Peter Caldwell at the Lawrence Livermore National Laboratory in California and his colleagues. Yet this is only one mechanism and there is still disagreement about whether stratocumulus will increase or decrease in a warming world.

That is because researchers have realized that several positive feedbacks could be at work. For one thing, the stratocumulus could be starved of moisture. Low clouds get their moisture by a roundabout process: as heat radiates from the cloud tops, cold parcels of air form and sink down. This pushes up warm, damp air from near the sea surface, which forms more cloud as it cools and condenses. In 2009 two teams that included Caldwell and Bjorn Stevens at the Max Planck Institute for Meteorology in Hamburg, Germany, pointed out that increased temperatures will reduce heat loss from the cloud tops. That would mean less cooling, less sinking air, less moisture dragged up and less cooling cloud cover on average.

And this is not the only potential positive feedback at work. Even where a temperature inversion traps the clouds, there is some mixing between damp, cool air below and dry, warm air above. One scenario suggested in 2012 is that warming could drive stronger mixing, dissipating the vital water. The result would be reduced cloud cover and amplified warming.

Even if mixing doesn't get stronger, there could still be more moisture loss to the dry air above. Warmer air can hold much more water vapour, so in a warmer world a given air current will carry away more moisture.

To find out how big the mixing effect is, Steven Sherwood at the University of New South Wales in Sydney, Australia, and colleagues looked at data from weather balloons. The mixing turned out to be pretty vigorous – more than in many climate models. Different models predict varying levels of climate sensitivity, the temperature increase resulting from a doubling of atmospheric CO_2. But the models with realistic mixing are the ones with greater sensitivity, Sherwood found. If these are to be trusted, then Earth's short-term sensitivity will be 3 °C to 4.5 °C.

Good work though this is, John Fasullo at the National Center for Atmospheric Research in Boulder, Colorado, reckons it is not definitive. The observations of mixing are limited – relying on a scattering of weather balloons – so it may be difficult to confirm the theory. Fasullo prefers to compare cloudiness directly with humidity, which can be measured globally by satellites. In 2012 he showed that models often overestimate the humidity in the subtropics. This finding is also bad news because the models with more realistic low humidity tended to predict greater warming.

So some light is being cast on the great cloud conundrum, but it is still rather a dingy grey light, just hinting at which models might be most trustworthy. As computer power increases we can build models with finer resolution, but we won't reach some paradise of perfect modelling. It will be several decades at least before global models can include the small-scale processes that take place in clouds. Until then, models must keep using approximations for this small-scale stuff.

Can geoengineering avert climate chaos?

Climate change is upon us, melting ice, fuelling storms and making floods and heatwaves more intense. Global emissions of carbon dioxide and other greenhouse gases continue to increase, promising worse to come. And even if we stopped all emissions tomorrow, some of these gases will stay stubbornly in the atmosphere, pushing up temperatures for decades.

Perhaps, then, it is time to get serious about the audacious idea of geoengineering. The hope is that by deliberately tinkering with our planet's climate machine, we might be able to fix our gargantuan blunder, or at least avoid some of the most serious consequences, or just buy ourselves a bit more time to cut emissions.

Dozens of schemes have been devised to cool the planet. We could launch a vast fleet of ships to whiten the clouds by spraying salt mist, or squirt sulfuric acid into the stratosphere to reflect sunlight. Send a swarm of mirrors into deep space. Engineer paler crops. Fertilize the oceans. Cover the world's deserts in shiny mylar. Spread cloud-seeding bacteria. Release a global flock of microballoons.

Ingenious these schemes may be, but would any of them work? Or would they just make things worse? Short of taking the biggest gamble imaginable and trying one out, the best we can do is try to explore each idea with detailed calculations and computer models. As the results of such studies mount up, we're starting to get an idea of what geoengineering might – or might not – be able to achieve.

Some ideas can be dismissed with relative ease. Covering deserts in reflective plastic, for example, could reflect a lot of sunlight and cool the planet, but it probably is as crazy as it sounds. It would devastate ecosystems, alter regional climate patterns and require an army of cleaners. Others are beyond our powers today. To shade Earth with a swarm of space parasols would require an estimated 20 million rocket launches. Without some radical new technology, that would be astronomically expensive and fatally polluting. Other schemes are certainly feasible – but can they actually fix the climate?

The basic problem, of course, is that rising levels of greenhouse gases in the atmosphere are trapping heat. Sometime this century we are likely to double the pre-industrial concentration of CO_2 in the atmosphere, which will reduce heat loss by about 3.7 watts per square metre, averaged across the planet. To stop Earth warming, any geoengineering scheme either has to block as much incoming heat from the Sun or increase heat loss from the top of the atmosphere by as much.

We have other prerequisites for our global refrigerator (see Figure 10.6). It needs to work without drastically altering regional climates, while also preventing sea level from rising. Ideally, we want to stop the oceans becoming so acidic that coral reefs vanish, too.

But the first test is potency. Tim Lenton of the University of Exeter and Nem Vaughan of the University of East Anglia in Norwich, both in the UK, combined various model results with their own calculations to assess the potential cooling power of a couple of dozen proposals. They found that many schemes would make little difference. Take the idea of making roofs and roads whiter to reflect more sunlight. Even with optimistic assumptions, this could only reflect about 0.15 watts per square metre – at best a minor contribution to restoring Earth's heat balance.

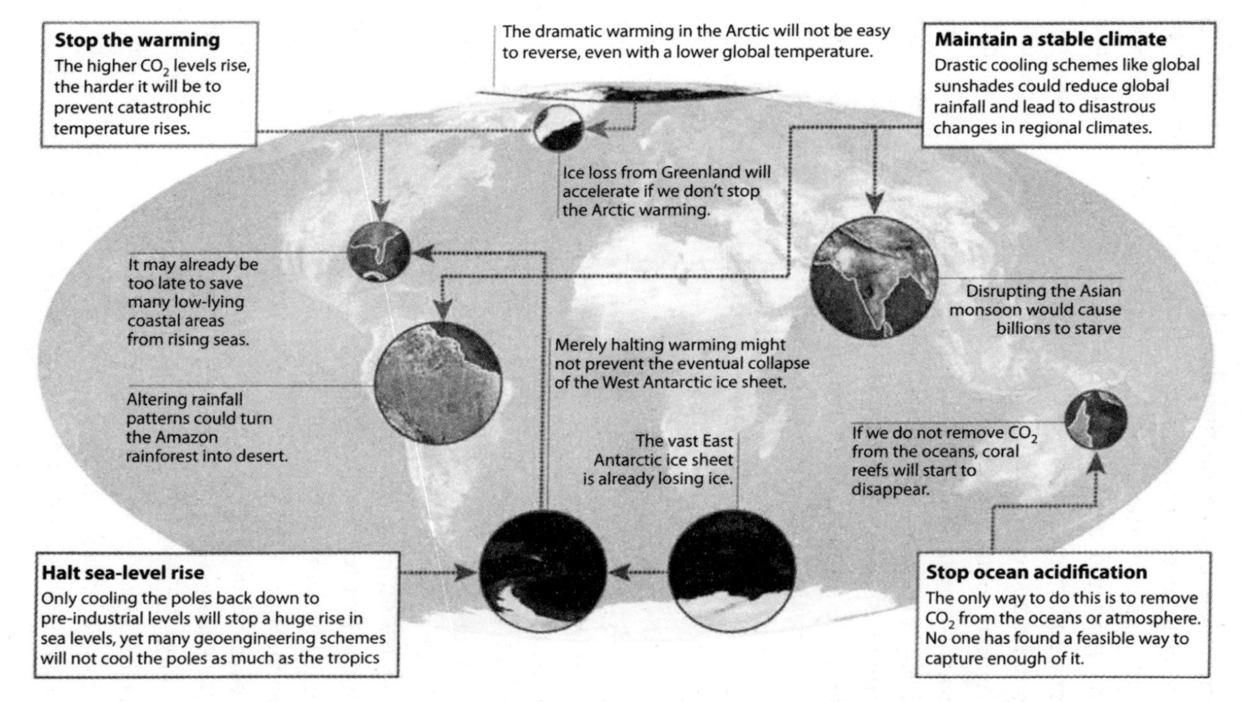

FIGURE 10.6 Cooling the planet is an immense task. But, to avert catastrophe, any geoengineering scheme must also meet other requirements.

What about fertilizing the seas? Phytoplankton consume CO_2 as they grow, and their dead bodies tend to sink to the sea floor and get buried, locking their carbon away. Adding nutrients to the oceans that are in short supply, such as iron, could boost phytoplankton growth. By the end of the century, this could improve the heat balance by as much as 0.2 watts per square metre, Lenton and Vaughan calculated. Again, this would not make enough of a difference.

Many other proposals, such as encouraging downwelling in polar regions to speeding up the transport of carbon into the ocean depths, are even more limited. But two schemes stand out as being both highly potent and relatively feasible. Both use some form of sunshade.

Putting global warming in the shade

One idea is to whiten marine clouds – specifically the low, flat stratus clouds that cover much of the tropical oceans. Ships around the globe would send plumes of fine salt spray up into the air. By acting as nucleation sites, the salt particles should encourage droplets of water to form in clouds. With more droplets per cubic metre, these clouds would be whiter than normal, and reflect more sunlight. Potentially, this could offset the entire warming from a doubling in CO_2.

Cloud-whitening has its upsides, such as not using hazardous chemicals. But cloud nucleation is not well understood (see Chapter 6), so it might not live up to expectations, and cooling only the oceans could disrupt local climates. A study published in 2012, for instance, found that seeding clouds over the Pacific might alter rainfall patterns in a similar way to El Niño's disruptive counterpart, La Niña.

The other leading contender is an old one: fill the atmosphere with a haze of fine particles. In fact, we have been doing it already. Sulfur dioxide pollution forms fine droplets of sulfuric acid that already reflect an estimated 0.4 watts per square metre. But SO_2 from fires and factories doesn't stay in the atmosphere for long, so its effects are limited. If sulfate reaches the stratosphere, however, it can linger for years and its cooling effect is much greater. The proof comes from volcanic eruptions such as Mount Pinatubo's in 1991, which cooled the planet by up to 0.5 °C over the following couple of years.

To balance the warming effect of a doubling in CO_2, we would need to pump up to 5 million tonnes a year of SO_2 into the stratosphere. According to Justin McClellan of Aurora Flight Sciences in Cambridge, Massachusetts, whose team evaluated several ways to deliver the sulfates, this would cost about $10 billion a year. Compared with the stupendous costs of global warming, this is an absolute bargain. Sea-level rise alone will swallow up many trillions of dollars' worth of cities and farmland.

The dark side of shade

Unfortunately, our sulfur spray may barely slow the seas' advance. Sulfur droplets do not linger as long in polar regions as they do in the tropics, making them less effective polar coolants. So even if aerosol injection brought the average global temperature down to that of the 1800s, the poles would still be warmer than they were and the ice sheets would keep melting.

It is not clear whether a different kind of reflector, such as solid metallic particles or tiny, shiny balloons, would do any better. Pumping out a gas is so much simpler and cheaper, so most studies have concentrated on sulfates.

While coastal plains and cities drown, the rest of the planet might dry out. Any kind of sunshade would reduce the light reaching the sea surface, reducing evaporation. So far, the effect of sulfur pollution has been outweighed by warming, which increases evaporation. But if we reduced the temperature to preindustrial levels this way, there would be a dramatic decline in rainfall. That might be avoided by not reducing the temperature as much – but then the ice sheets would melt faster.

Sunshades could also have disastrous regional effects, according to climate models. If they disrupted the monsoons, they could bring permanent famine to billions. Or changing the circulation patterns that feed moisture to the Amazon basin could turn it into a desert.

Myles Allen of the University of Oxford, UK, and his colleagues looked at the effect of varying amounts of sunscreen in the stratosphere using a detailed climate model. They found that there is no solution that works for everyone. An amount of aerosol that would take China close to comfortable preindustrial temperature and rainfall might cool India far too much.

Or it could be the other way around. Climate models agree fairly well on the global effects of sunshade schemes, but produce different regional impacts. This may be because of the different assumptions and values used in different studies. Or it may be due to the limitations of existing climate models.

Even if they were all accurate, some factors affecting regional climates are inherently unpredictable. How will ecosystems respond, for example? As a result, we can never be 100-per-cent certain that any particular scheme will have the desired result.

This makes any sunshade highly risky. If it turned out to have some terrible consequence and we suddenly stopped replenishing sulfates or whitening clouds, the planet would warm very rapidly over the next few years. Such a sudden transition would be even more damaging than a gradual warming to the same level, giving no time for people and wildlife to adapt. If we reach for the sulfates, we might need another type of geoengineering, such as cirrus seeding to cool the poles, prescribing not just one but two dangerous drugs for the planet.

You cannot be cirrus

The high, wispy cirrus clouds that sometimes grace an otherwise blue summer sky may seem an unlikely enemy, but David Mitchell has plans to attack them. Destroying cirrus might not only reduce global temperature but also help save the ice caps and curb extreme weather.

Icy cirrus clouds reflect and radiate much less heat into space than lower cloud, so their net effect is to warm us up. In 2009, Mitchell – based at the Desert Research Institute in Reno, Nevada – suggested that we could use aircraft to spread bismuth triiodide, a non-toxic compound that should seed relatively large ice crystals. These would fall from the sky faster than natural cirrus ice, so the clouds would disperse.

Preliminary attempts to model the process, indicated that this could cool the planet by about 2 watts per square metre – enough to prevent half of the warming from a doubling of CO_2.

Better still, the method ought to work best where it is most needed, at high latitudes. Concentrating efforts here could protect our fragile ice caps. It would also help to restore the temperature difference between tropics and poles.

Mitchell cautions that lots of research needs to be done on representing cirrus in global climate models, and not just for geoengineering. He would like to see a cloud-seeding experiment in a small area to see what really happens. What's more, dispersing cirrus shares many of the risks of sunshade schemes: it may well have disastrous regional effects, while stopping abruptly would be dangerous.

Don't shade, scrub

Instead of blocking sunlight, maybe we should get at the real cause of the problem and actively scrub CO_2 from the air. The concentrated gas could then be pumped into underground reservoirs such as depleted gas and oil fields. But no one has so far devised an efficient method for doing this. The problem, according to Lenton, stems from trying to capture a very dilute gas, which is inherently costly compared with capturing CO_2 from a concentrated source like power-station smoke.

With existing technology, there is no realistic prospect of mopping up all the extra CO_2 we are adding to the atmosphere in time to prevent further climate change. Instead of covering the planet in carbon-eating machinery, how about speeding up the reaction of CO_2 with silicate rocks? Over millions of years, this process, called chemical weathering, has soaked up vast amounts of CO_2 (see Chapter 6). But to deal with just a single year's worth of emissions, we'd need to grind up at least 7 cubic kilometres of rock and spread it so thinly that it would cover several per cent of Earth's land surface. So this process cannot save us, either.

What about modifying land use and agriculture to capture more carbon? Simply planting forests remains a good thing, although geography limits its potential to about 0.5 watts per square metre, and all that carbon could end up back in the atmosphere if forests die or burn as the planet warms.

One way to lock away the carbon stored in plants is to turn it into charcoal – biochar – and bury it. Another is to burn crops in power plants fitted with carbon-capture technology. These ideas need land, so they will compete with food production. Lenton calculated that the total benefit could be just 0.3 watts per square metre by 2050.

In the end, the greatest obstacle to any drastic form of geoengineering may turn out to be politics. Allen points out that you can't have competing geoengineering programmes. There would have to be one supranational body deciding to go or no-go. Achieving agreement may be almost impossible, because different countries will have different priorities. Some are most threatened by sea-level rise, others by sheer heat or shifting rainfall.

Where does all this leave us? It is true that international agreement will be needed only for big sunshield schemes, with their global dangers; individuals, institutions or countries could act unilaterally when, say, using biochar or planting a forest. But, as Lenton found, these methods are simply not up to the task. A 2018 report by the European Academies Science Advisory Council found that such carbon-capture schemes, including ocean fertilization, have only 'limited realistic potential' to have a meaningful effect on greenhouse gas levels. The best strategy, the Council concludes, is to stop those gases getting into the atmosphere in the first place.

11
Conclusion

How do you end a book about planet Earth? By discussing the planet's end, of course. Barring some unforeseen catastrophe, Earth's nemesis is likely to be the star that has made it fit for life for billions of years. Our sun is not destined to explode as a supernova, hurling its planets into space. It's just not massive enough. But when it finally burns through its central supply of hydrogen some 6 billion years from now, it will inflate spectacularly and consume its nearest neighbours.

The Sun shines today thanks to nuclear fusion in its core, which generates energy by converting hydrogen to helium. Once all the hydrogen there has been consumed, fusion will ignite a shell of hydrogen around the core. The extra energy this generates will make the Sun thousands of times more luminous than it is now and overpower its own gravity so that it will engulf the inner planets. At full splendour, its radius will extend beyond Earth's present orbit.

There is just a chance that our little blue marble may escape. As the Sun swells, it will slough off perhaps a third of its mass in a great outrush of charged particles. Its reduced gravitational pull will allow its captive planets to expand their orbits.

So it will be a race to outrun the Sun as it grows. Mercury, then Venus, will almost certainly lose, each being engulfed by the Sun's inflated atmosphere. The fate of Earth is too close to call, according to Dimitri Veras at the University of Warwick,

UK. Still, Earth will not escape completely: fierce tides from the Sun's outer layers will cook the planet's interior, generating a serious breakout of volcanism.

Of course, by this time, life on Earth is likely to be a mere memory – or very different from what we know now. That's because there are at least two killer crises heading our way. Firstly, in at most a billion years from now, the naturally increasing heat from the Sun will drive the planet's thermostat into overdrive, stripping more and more CO_2 out of the atmosphere until there is not enough to support photosynthesis. Ultimately, all plants will die and all life that depends on plants will fail (see Chapter 8).

If we find a way to survive this – and it's a big if – in some 3 to 4 billion years continued cooling of Earth will see the entire core freeze solid. When this happens, the engine that drives plate tectonics will stop and volcanoes will go extinct. Earthquakes will continue for a while as the crust cools and thickens. But the Earth will probably no longer be hospitable to life. With no molten core to maintain its magnetic field, the planet will be ravaged by the solar wind.

These are natural calamities currently outside our control. But another disaster is facing us right now which is of our own making. Unless we stop emitting CO_2 and other greenhouse gases, or find realistic ways to remove them from the atmosphere, global warming is soon going to threaten life on a frightening scale (see Chapter 10).

This is a gloomy conclusion indeed. But there are good reasons to be hopeful. We humans are ingenious. And, evolutionarily, we are odd: we help others, even complete strangers. Few other species do that. At a global level, this unusual cooperation enabled us to start sealing the ozone hole over Antarctica (see Chapter 6). Let us hope national governments can repeat this success and make meaningful cuts in levels of greenhouse gases in the atmosphere. That way, humans could still be thriving in a billion years from now – and by then we may even have set up a colony on our new planet, Earth II.

50 down-to-Earth ideas

This section helps you to get to know our planet in greater depth, with more than just the usual reading list.

Ten revealing sites to visit

1 **Iceland** is one big brilliant geology exhibit. It includes Geysir, the first geyser to be described in print, and Grímsvötn, the island's most active volcano. For things to do, try a tour through glacier ice caves or inside the dormant Thrihnukagigur volcano, bathe in a hot spring pool in all weathers or smear thermal mud on your face. Iceland was created by the upwelling of magma as the North American and Eurasian plates began drifting apart. At Thingvellir, you can walk between these two tectonic plates. The island also has a new museum at https://lavacentre.is

2 **Lanzarote** in the Canary Islands is a great place for an introduction to volcanism. Timanfaya National Park covers a large part of the island, which was flooded by lava in the eighteenth and nineteenth centuries. The island's meagre rainfall means erosion is low and the area looks pretty much as it did after the eruptions. If you fancy it, you can have a meal cooked over hot lava. To work up an appetite, there are plenty of walks around calderas and through the Martian-like lava fields. The visitor centre in Mancha Blanca gives a good overview of Earth's structure, plate tectonics and volcanoes (http://www.lanzaroteinformation.com/content/timanfaya-vistors-centre).

3 On another island half a world away, the **Hawai'i Volcanoes National Park** (https://www.nps.gov/havo/) is a real hotspot. The rangers are knowledgeable and the entire park is built on the premise of letting people get as close as safely possible to the volcanoes, lava flows

and sulfur vents. The Thomas A. Jaggar Museum, which overlooks Kīlauea Caldera, has amazing views.

4 Staying in the Pacific, **New Zealand** is also a hotbed of volcanism. The islands are located where the Pacific plate is subducting under the Indo-Australian plate. The North Island has national parks built around volcanoes and areas of geothermal activity. You can visit the excavated Maori village of Te Wairoa, which was destroyed in 1886 by the eruption of Mount Tarawera. Frying Pan Lake is perhaps the world's largest hot lake and White Island in the Bay of Plenty is home to the country's most active cone volcano; it has been smoking since at least 1769, when Captain Cook first spied it. For a view of a country living with volcanoes, try the Auckland Museum (http://www.aucklandmuseum.com/volcanoes).

5 Talking of museums, London's **Natural History Museum** is a great place to discover new things about Earth. You'll find plenty of fossils and exhibitions devoted to the planet. You can even experience what it's like to be in an earthquake with the museum's quake simulator (www.nhm.ac.uk).

6 **Dynamic Earth** in Edinburgh, Scotland, is a very modern take on presenting the story of Earth. You can feel the ground shudder as molten lava speeds towards you, watch volcanoes throw ash into the air and see the Northern Lights dance above the ice caps. If you don't like being penned in, take a walk up Arthur's Seat next door – a remnant of a volcanic eruption in the Carboniferous (http://www.dynamicearth.co.uk/).

7 Marking England's southern boundary is the **Jurassic Coast**, an in-situ collection of sedimentary rocks and fossils that document 185 million years of Earth's Triassic, Jurassic and Cretaceous past. Its golden beaches, sheltered coves and sculpted cliffs, stacks and arches stretch for 150 kilometres, and it is a prime destination for fossil hunters. A good place to start is the Lyme Regis Museum (http://www.lymeregismuseum.co.uk).

8 A slightly larger collection of Earthly goodies than Lyme Regis's can be found in Beijing, China, in the **Geological Museum of China**. It's home to 200,000 objects including the world's largest quartz crystal, weighing in at 3.5 tonnes, and fossils of the giant Shangdong dinosaur and rare primitive birds (http://chinamuseums.com/geological.htm).

9 If you want to see a rip in Earth's crust, why not visit the **San Andreas Fault**, which runs for 1200 kilometres roughly following California's coast. It marks the boundary between the north-moving Pacific plate and the North American plate, which is heading south. The tear becomes visibly impressive at a variety of places along its length, such as on the crumpled grasslands of Carrizo Plain National Monument and in Pinnacles National Park. There are plenty of organized tours, or make up your own at http://www.sanandreasfault.org/.

10 For drama and excitement – and funds permitting – you may like to try storm-chasing across the **Great Plains tornado belt** in the USA. It is one way to see

the troposphere in turmoil and judge how good we are at forecasting weather. Through May and June you can buy a seat on a tour that will search out tornados and supercell thunderstorms.

Ten of Earth's biggest bangs

1 One of the twentieth century's most notorious bangs happened at 7.14 a.m. on 30 June 1908. At that moment, something exploded with enormous force over the Tunguska River in Siberia, Russia. The shockwave flattened trees over an area of 2000 square kilometres, and people tens of kilometres away were knocked off their feet. The blast, which has been estimated as the equivalent of 15 megatons of TNT, is widely thought to have been caused by a meteor a few tens of metres across exploding in mid-air. But debate continues and other possibilities, such as underground methane, are still being investigated. If it was a meteor, it was a mere tiddler by historic standards.

2 The biggest human-made bang of all time was the detonation in October 1961 of the Tsar Bomba. The Soviet-built hydrogen bomb blew up over Novaya Zemlya off the north coast of Russia with a flash of light that was visible 1000 kilometres away. Windows shattered 900 kilometres away, and the mushroom cloud rose 64 kilometres above ground. At around 57 megatons, it released four times the energy of the Tunguska meteorite. After the test, a visitor to the blast site said, 'The ground surface of the island has been levelled, swept and licked so that it looks like a skating rink.'

3 One of the most famous explosive events in recent history was the 1883 eruption of Krakatoa, a volcanic island in the Sunda Strait between the Indonesian islands of Java and Sumatra. Following months of escalating activity, the volcano erupted in late August. On

the most violent day, 27 August, there were four enormous explosions, audible up to 4000 kilometres away. Tens of thousands were killed and many more injured. Krakatoa itself was virtually destroyed, although later eruptions have replaced it with a new island.

At this point we have to switch scales. Eruptions are measured on the Volcanic Explosivity Index (VEI), which goes from 0 (weakest) to 8 (strongest). Each step on the scale represents a tenfold increase in the volume of rock and ash disgorged. Krakatoa, rated 6, was equivalent to about 200 megatons of TNT – at least three times the energy of Tsar Bomba.

4 Only one eruption in recent history has made it to 7 on the Volcanic Explosivity Index – ten times the size of Krakatoa. That was Mount Tambora, on the island of Sumbawa in Indonesia. Tambora started to grumble in 1812, building up to a cataclysmic eruption in April 1815. The outburst flung vast amounts of dust and ash into the atmosphere, cooling temperatures around the globe. The following year – 1816 – became known as the 'year without a summer'. Amazing as it now sounds, Tambora was largely overlooked for some 160 years. Only when scientists examined ash layers in Greenland ice cores was the truth uncovered: no volcanic eruption since 1815 has amounted to more than a damp squib compared with Tambora.

5 Today New Zealand's Lake Taupo is a serene body of fresh water, beloved of crayfish and hikers alike. But 26,500 years ago, it was the site of a massive volcanic blast that coated North Island with ash and rock 200 metres deep. The present-day lake lies in the caldera of

the volcano. What's called the Oruanui eruption was the most recent volcanic event to score the maximum 8 on the VEI – ten times the scale of Tambora.

6 Volcanologists today like to talk about eruptions that far outstrip those of common-or-garden volcanoes. Such 'supervolcanoes' would include eruptions that release more than 1000 cubic kilometres of material. The Oruanui eruption probably managed that, but a surer candidate is Sumatra's Toba blast, which released 2800 cubic kilometres of hot rock, ash and dust about 75,000 years ago. It formed an enormous caldera, now partly filled by a lake. The vast amount of ash is thought by some to have significantly chilled the world and caused a collapse in human population, an idea that is still controversial.

7 Beneath the beauty of Yellowstone National Park lies a monster. A plume of hot molten rock that rises from deep within Earth is believed to cause massive eruptions periodically, some of them big enough to be termed supervolcanoes. The Yellowstone hotspot has gone off three times in the past few million years. The Huckleberry Ridge eruption 2.1 million years ago – which was almost as large as the Toba blast – formed the Island Park caldera. A smaller, but nonetheless still supervolcanic, eruption formed the Henry's Fork caldera 1.3 million years ago, and the Lava Creek eruption 640,000 years ago formed the present-day Yellowstone caldera. The first and third eruptions covered most of North America with ash.

8 Possibly the largest single volcanic eruption in Earth's history, La Garita blasted 5000 cubic kilometres of

material on to Earth's surface around 27 million years ago. A huge area was devastated, and much of the resulting volcanic rock can still be seen in what is now Colorado. The eruption seems to have started with the volcano spitting out lumps of rock up to 2 metres across, before releasing an enormous pyroclastic flow – a fast-moving gush of hot gas and rock. Fortunately for us, the La Garita supervolcano is now extinct.

9 For bangs bigger than volcanoes can offer, we have to return to cosmic collisions. Beneath the town of Chicxulub on Mexico's Yucatán Peninsula, buried under vast amounts of sediment, is a crater 180 kilometres across. It was made by an asteroid about 10 kilometres across that slammed into the planet 65 million years ago – the biggest meteoric impact in the past billion years, with an explosive force of around 100 million megatons. Most palaeontologists accept that this impact was at least partly responsible for the extinction of the dinosaurs 65 million years ago, though some still disagree. The 'killer' would not have been the actual explosion, but the ecological devastation wrought by the resulting dust cloud. In this, the asteroid may have been aided by massive volcanic outpourings in India, in the area known today as the Deccan traps.

10 Perhaps the 'big splat' that threw up the Moon should be here, but we discussed that in Chapter 1. So, instead, let's look forwards…

Unfortunately for us, it's unlikely that Earth-shaking explosions are a thing of the past. The world's nuclear arsenal includes some 15,000 weapons whose total destructive power amounts to thousands of megatons.

The likelihood of these going off has fallen since the end of the Cold War, but remains high.

As well as a plethora of 'normal' volcanoes, six supervolcanoes have been identified: Taupo, Toba and Yellowstone, plus the Valles caldera in New Mexico, Long Valley in California and the Aira caldera in Kagoshima Bay, Japan. We don't know if or when any of them will explode again, though our luck probably won't hold out for ever.

As for cosmic collisions, most of the half-million known asteroids and 12 trillion comets will never come near Earth. Only two asteroids are rated as having any chance of hitting our planet at all: one in 2048 and the other in 2880. But with both rated at just 1 on the Torino impact hazard scale, which goes from 0 to 10, it's unlikely we'll go the way of the dinosaurs – yet.

Five superlatives

We all know the highest mountain and deepest ocean trench, but here are a few record-breakers you may not have heard of.

1. The world's greatest concentration of geysers exists in Upper Geyser Basin in Yellowstone National Park, Wyoming, USA. It boasts 150 hot jets in 2.6 square kilometres, including Old Faithful.

2. The highest temperature ever recorded was 56.7 °C, in Death Valley, California, USA on 10 July 1913. For 90 years the record-holder was El Aziza in Libya, which hit 58 °C in 1922. But in 2012 the World Meteorological Organization disqualified the reading over doubts about its accuracy.

3. The longest cave system is the limestone labyrinth at Mammoth Cave in Kentucky, USA. It stretches for 663 kilometres and is twice as long as its nearest rival.

4. Vredefort in South Africa is home to the world's largest impact crater. It measured 300 kilometres in diameter when it was created around 2 billion years ago. The culprit was an asteroid of between 10 and 15 kilometres across. Sadly, only remnants of the original crater can be seen today.

5. The Dry Valleys near McMurdo Sound in Antarctica are well named. They have received no rain or snow for 2 million years, making them the driest places on Earth. They form an extreme desert, drier even than the Atacama Desert of Chile and Peru.

Five Earthy jokes

1 A: What should I tell my brother about marrying a geologist?

B: Tell him that geologists have their faults, and that the more he tries to be gneiss, the more he'll be taken for granite. And warn him on his wedding night that geologists make the bedrock!

(Granite you know. Gneiss is a metamorphic rock.)

2 You may be a geologist if:

- You can pronounce the word 'molybdenite' correctly on the first try.
- You own more pieces of quartz than underwear.
- Baggage handlers at the airport know you by name and refuse to help with your luggage.
- You watch Western movies just for the rock formations.
- Your children are named Jewel, Rocky and Beryl.
- You can point out where Tsumeb is on a map.
- You shouted 'obsidian!' to cinemagoers when watching *The Shawshank Redemption*.

(Tsumeb is a town in Namibia famous for its mine, which has yielded some of the largest crystals of dioptase, an intense green mineral. The hero of *The Shawshank Redemption* leaves a message under a piece of black volcanic glass – known to geologists as obsidian.)

3 Hurricane Harvey and El Niño are having a drink in a bar. 'I'm so tough,' boasts Harvey, 'I can devastate entire island economies and cause multimillion dollar damage to Florida.'

'That's *nada*,' says El Niño dismissively. 'I can cause flooding in deserts, the desiccation of rainforests. Entire ecosystems thrive or die at my mere whim. The economies of nations are subject to my vacillations.'

Upon which a small North Atlantic low-pressure system enters the bar, precipitating meekly on the floor. Hurricane Harvey and El Niño dive for cover behind the bar, trembling.

'What's up with you?' jeers the barman, 'I thought you two were the toughest meteorological phenomena in town!' 'We're tough,' wails El Niño piteously, 'but he's cyclonic!'

4 Global warming is so funny that the ice sheets are cracking up!

5 Two planets meet. The first one asks: 'How are you?' 'Not so well,' the second answers. 'I've got *Homo sapiens*.' 'Don't worry,' the other replies, 'I had the same. It won't last long.'

Ten quotes

1 'But the whole vital process of the Earth takes place so gradually and in periods of time which are so immense compared with the length of our life, that these changes are not observed, and before their course can be recorded from beginning to end whole nations perish and are destroyed.' (Aristotle, Greek philosopher)

2 'Daily it is forced home on the mind of the geologist that nothing, not even the wind that blows, is so unstable as the level of the crust of this Earth.' (Charles Darwin, co-originator of evolutionary theory)

3 'The forces which displace continents are the same as those which produce great fold-mountain ranges. Continental drift, faults and compressions, earthquakes, volcanicity, transgression cycles and polar wandering are undoubtedly connected causally on a grand scale. Their common intensification in certain periods of the earth's history shows this to be true. However, what is cause and what effect, only the future will unveil.' (Alfred Wegener in prophetic mood. He was a German geophysicist who proposed the idea that the continents moved but could find no force to drive them.)

4 'India in the Oligocene crashed head on into Tibet, hit so hard that it not only folded and buckled the plate boundaries but also plowed into the newly created Tibetan plateau and drove the Himalayas five and a half miles into the sky… When the climbers in 1953 planted their flags on the highest mountain, they set them in snow over the skeletons of creatures that had

lived in a warm clear ocean that India, moving north, blanked out. Possibly as much as 20,000 feet below the sea floor, the skeletal remains had turned into rock. This one fact is a treatise in itself on the movements of the surface of Earth.

'If by some fiat, I had to restrict all this writing to one sentence; this is the one I would choose: the summit of Mount Everest is marine limestone.' (John McPhee, US writer, author of *Annals of the Former World*)

5 'No geologist worth anything is permanently bound to a desk or laboratory, but the charming notion that true science can only be based on unbiased observation of nature in the raw is mythology. Creative work, in geology and anywhere else, is interaction and synthesis: half-baked ideas from a bar room, rocks in the field, chains of thought from lonely walks, numbers squeezed from rocks in a laboratory, numbers from a calculator riveted to a desk, fancy equipment usually malfunctioning on expensive ships, cheap equipment in the human cranium, arguments before a road cut." (Stephen Jay Gould, American palaeontologist, evolutionary biologist and writer)

6 'Carbon dioxide has not been at such a high atmospheric concentration for 3 million years. We understand its greenhouse effect – warming is unequivocal.' (Joanna Haigh, British atmospheric physicist)

7 'We've lost about 75 per cent of the Arctic's sea ice in just 30 years. It doesn't take a scientist to see this huge change in the climate system, and we know it's mainly due to the extra heat trapped by increasing

greenhouse gases from burning fossil fuels. Watching the Arctic warm so fast gives me the chills!' (Jennifer Francis, American atmospheric scientist specializing in the Arctic)

8 'What's the use of having developed a science well enough to make predictions if, in the end, all we're willing to do is stand around and wait for them to come true?' (F. Sherwood Rowland, American atmospheric chemist and Nobel laureate)

9 'The sea is the universal sewer where all kinds of pollution end up conveyed by rain from the atmosphere and from the mainland.' (Jacques Cousteau, marine explorer, conservationist, film-maker and co-developer of the aqua-lung)

10 'Every credible scientist on earth says your products harm the environment. I recommend paying weasels to write articles casting doubt on the data. Then eat the wrong kind of foods and hope you die before the Earth does.' (Dogbert advising his boss in a *Dilbert* cartoon strip written by Scott Adams, American cartoonist, humourist… and philosopher)

Ten ideas for further reading

1 *The Earth: An Intimate History* (2005) is Richard Fortey's romp through deep time, in which he relates everything about natural history, our culture and the creation of our cities to geology.

2 *Annals of the Former World* (2000) by John McPhee paints a vivid picture of North America's geology and is beautifully written. In fact, it won the 1999 Pulitzer Prize for Nonfiction.

3 Earth changes only at a slow snail's pace over geological time, right? No, says Michael R. Rampino in his book *Cataclysms: A New Geology for the Twenty-first Century* (2017). Rampino focuses on the drama of such things as asteroid impacts and gigantic outpourings of lava as rapid agents of change on our planet.

4 Britain's early practitioners of geology were in large part wealthy men less interested in science than in finding coal seams and scientific evidence of biblical events. In *Reading the Rocks* (2017), Brenda Maddox introduces the 'gentlemen geologists' (and a few women) of the eighteenth and nineteenth centuries. Their legacy was to reduce the Church's power and to help create secular science.

5 If fossils are your thing, then *Written in Stone: Evolution, the Fossil Record, and Our Place in Nature* (2011) by Brian Switek is worth a look. He delves deep into the fossil record and focuses on the 'missing links' – transitional fossils – which sit between one group of extinct

creatures and their descendants. Expect tales of walking whales, fish with feet and feathered dinosaurs.

6 In *The Great Quake: How the Biggest Earthquake in North America Changed Our Understanding of the Planet* (2017), Henry Fountain dissects the 1964 Alaska earthquake and examines its wider scientific impact. Plate tectonics was being developed in the mid-1960s and Fountain uses this broader context as the background for his tale.

7 *Oceans: A Very Short Introduction* (2017) by Dorrik Stow lets you dip a toe into the subject. And, having mentioned Oxford University Press's Very Short Introduction series, it is worth noting that it includes lots of terrestrial topics from Earth to atmosphere, minerals to plate tectonics.

8 For an enjoyable journey, try Gabrielle Walker's *An Ocean of Air: A Natural History of the Atmosphere* (2008). You'll learn about the differing natures of the atmosphere's onion-like layers, their component gases and the people who shaped our understanding of them.

9 The notion of Mother Earth, 'Gaia', as a self-nurturing superorganism may be losing scientific credibility but its accompanying message – that we must treat our planet better or risk disaster – has lost none of its importance. James Lovelock is worth reading for this reason alone: he has been a major force in shaping modern views of Earth. His latest book on the theme is *The Vanishing Face of Gaia: A Final Warning* (2009).

10 Ecologist George Woodwell's latest offering warns us about the impact of climate change on the plant and animal communities that maintain the biosphere on which we are so dependent. You can read his solution in *A World to Live In: An Ecologist's Vision for a Plundered Planet* (2016).

Glossary

Aerosol A suspension of fine solid or liquid particles in air. Some aerosols act as ice nucleators in clouds, others reflect solar radiation back into space.

Asthenosphere The mechanically weak, highly viscous layer of the mantle below the lithosphere.

Chondrites Stony meteorites that formed at the same time as our planet and match many aspects, though not all, of Earth's composition.

Climate sensitivity The increase in Earth's average surface temperature in response to a doubling of atmospheric carbon dioxide concentration.

El Niño Warm phase of the El Niño Southern Oscillation (ENSO), in which sea surface temperature rises in the tropical central and eastern Pacific. It is highly disruptive to regional and global weather patterns.

Geologic timescale The scale that puts a chronological date to the different layers of rock that have been laid down over time. It is derived from the stratigraphic scale.

Geoneutrino Antimatter variant of the neutrino – an electron antineutrino – which is generated by the decay of radioactive isotopes within Earth.

Igneous rock Rock formed when magma or lava cools and solidifies.

Isotopes Variants of a particular element, differing in the number of neutrons in their nuclei. Some are radioactive. Isotopes are widely used in dating rock samples.

La Niña Cool phase of ENSO, in which sea surface temperature falls in the tropical central and eastern Pacific. Like its counterpart, El Niño, it can severely disrupt weather around the world.

Lava Molten rock that has reached Earth's surface.

Lithosphere Outer, solid shell of Earth. It includes the crust and rigid upper mantle.

Magma Molten rock enclosed within the planet. It may collect in chambers beneath volcanoes and, once ejected, becomes lava.

Mantle The largest layer within Earth between the outer core and crust.

Metamorphic rock Rock formed from older rocks that have been exposed to extreme temperature and pressure within Earth, causing profound physical or chemical changes.

Mineral A naturally occurring chemical compound. Rocks are made of one or more minerals.

Neutrino Electrically neutral, near-massless subatomic particle that interacts only weakly with normal matter.

Olivine The mineral magnesium iron silicate is a major constituent of the upper mantle. It is found in igneous rocks, such as basalt and undergoes rapid chemical weathering to form magnesium carbonate, locking up carbon from the atmosphere.

Ophiolite A section of ancient oceanic crust that, instead of being drawn into Earth by subduction, is raised up. It often sits on top of less dense continental crust.

Pyroxenes A group of silicate minerals found in the mantle and igneous rocks, containing calcium, sodium, magnesium, iron or aluminium.

Sedimentary rock Rock made by the compression of fragments of older rocks and organic detritus that has collected on river, lake or sea beds. It usually forms in layers (strata).

Solar wind A stream of high-energy charged particles produced by the Sun.

Stratigraphic scale The order in which different layers, or strata, of rock were laid down over time. Otherwise called the geologic record, it forms the basis for the geologic timescale.

Subduction The process that draws old oceanic crust down into the mantle. It takes place where one tectonic plate converges on another and is thought to be the main driver of plate movements.

Zircon A mineral, zirconium silicate, which forms tiny, resilient crystals that can contain traces of other minerals. Some are among the oldest objects on the planet and are used to deduce information about early Earth.

Picture credits

Figure 1.1 NASA/Goddard Space Flight Center/Arizona State University

Figure 1.2 Birger Rasmussen/Curtin University

Figure 2.1 imageBROKER/SuperStock

Figure 2.3 B&C Alexander/ArcticPhoto

Figure 2.4 Karsten Wrobel/imageBROKER/REX/Shutterstock

Figure 3.3 KamLAND

Figure 3.3 Ragnar Th. Sigurdsson/age fotostock/SuperStock

Figure 4.2 American Geophysical Union (AGU), courtesy AIP Emilio Segre Visual Archives

Figure 4.5 MUSTAFA OZER/AFP/Getty

Figure 5.1 Tom Till/SuperStock

Figure 5.3 G. Brad Lewis/Aurora/SuperStock

Figure 6.5 Mike Hollingshead/Solent News/REX/Shutterstock

Figure 6.6 Dan Leffel/age fotostock/SuperStock

Figure 6.8 Craig Eccles/Solent News/REX/Shutterstock

Figure 7.3 John Lund/Stone/Getty

Figure 7.5 OguzMeric/iStock/Getty Images Plus

Figure 8.1 Kerstin Langenberger/imageBROKER/SuperStock

Figure 8.3 Colin McPherson/Corbis via Getty Images

Figure 9.1 REX/Shutterstock

Figure 9.3 Paulo Oliveira/Alamy

Figure 10.1 Gideon Mendel For Action Aid/In Pictures/ Corbis via Getty

Figure 10.4 Tong Hong Wai/EyeEm/Getty

Index